T0407059

Bicycle Urbanism

Over recent decades, bicycling has received renewed interest as a means of improving transportation through crowded cities, improving personal health, and reducing environmental impacts associated with travel. Much of the discussion surrounding cycling has focused on bicycle facility design—how to best repurpose road infrastructure to accommodate bicycling. While part of the discussion has touched on culture, such as how to make bicycling a larger part of daily life, city design and planning have been sorely missing from consideration.

While interdisciplinary in its scope, this book takes a primarily planning approach to examining active transportation, and especially bicycling, in urban areas. The volume examines the land use aspects of the city—not just the streetscape. Illustrated using a range of case studies from the USA, Canada, and Australia, the volume provides a comprehensive overview of key topics of concern around cycling in the city including: imagining the future of bicycle-friendly cities; integrating bicycling into urban planning and design; the effects of bike use on health and environment; policies for developing bicycle infrastructure and programs; best practices in bicycle facility design and implementation; advances in technology, and economic contributions.

Rachel Berney is an Assistant Professor in the Department of Urban Design and Planning at the University of Washington, Seattle, USA.

Urban Planning and Environment

Series Editors: Donald Miller and Nicole Gurran

Maintaining and enhancing living conditions in cities through a combination of physical planning and environmental management is a newly emerging focus of governments around the world. For example, local governments seek to insulate sensitive land uses such as residential areas from environmentally intrusive activities such as major transport facilities and manufacturing. Regional governments protect water quality and natural habitat by enforcing pollution controls and regulating the location of growth. Some national governments fund acquisition of strategically important sites, facilitate the renewal of brownfields, and even develop integrated environmental quality plans. The aim of this series is to share information on experiments and best practices of governments at several levels. These empirically based studies present and critically assess a variety of initiatives to improve environmental quality. Although institutional and cultural contexts vary, lessons from one commonly can provide useful ideas to other communities. Each of the contributions are independently peer reviewed, and are intended to be helpful to professional planners and environmental managers, elected officials, representatives of NGOs, and researchers seeking improved ways to resolve environmental problems in urban areas and to foster sustainable urban development.

A full list of titles in this series is available at: www.routledge.com/Urban-Planning-and-Environment/book-series/UPE.

Recently published titles:

Bicycle Urbanism
Reimagining Bicycle Friendly Cities

Edited by
Rachel Berney

Routledge
Taylor & Francis Group

LONDON AND NEW YORK

First published 2018
by Routledge
2 Park Square, Milton Park, Abingdon, Oxon OX14 4RN

and by Routledge
711 Third Avenue, New York, NY 10017

Routledge is an imprint of the Taylor & Francis Group, an informa business

British Library Cataloguing-in-Publication Data
A catalogue record for this book is available from the British Library

Library of Congress Cataloging-in-Publication Data
Names: Berney, Rachel, editor.
Title: Bicycle urbanism : reimagining bicycle friendly cities / edited by Rachel Berney.
Description: Abingdon, Oxon ; New York, NY : Routledge, 2018. |
Series: Urban planning and environment series |
Includes bibliographical references and index.
Identifiers: LCCN 2017037280| ISBN 9781472456632 (hardback) |
ISBN 9781315569338 (ebook)
Subjects: LCSH: Bicycle commuting. | Urban transportation. |
City planning. | Bicycle lanes—Planning.
Classification: LCC HE5736 .B527 2018 | DDC 388.3/472—dc23
LC record available at https://lccn.loc.gov/2017037280

ISBN: 978-1-4724-5663-2 (hbk)
ISBN: 978-1-315-56933-8 (ebk)

Typeset in Sabon
by Florence Production Ltd, Stoodleigh, Devon, UK

MIX
Paper from
responsible sources
FSC
www.fsc.org   FSC® C013056

Printed and bound in Great Britain by
TJ International Ltd, Padstow, Cornwall

The right to the city is far more than the individual liberty to access urban resources: it is a right to change ourselves by changing the city. It is, moreover, a common rather than an individual right since this transformation inevitably depends upon the exercise of a collective power to reshape the processes of urbanization.

David Harvey, "The Right to the City"

Contents

Contributors

Rachel Berney is an Assistant Professor in the Department of Urban Design and Planning at the University of Washington, Seattle where she researches and teaches on public space, non-motorized transportation and mobility, and community design and development. Dr. Berney received her BLA from the University of Washington, Seattle and her MCP and PhD from the University of California, Berkeley. She is the author of *Learning from Bogotá: Pedagogical Urbanism and the Reshaping of Public Space*.

Derek Chisholm is the AECOM Transportation Operations Manager for the Gulf Coast, and is the Complete Streets Practice Lead for the firm. He has led numerous transportation plans and projects throughout the United States and abroad. Mr. Chisholm has served as a Project Manager, Adjunct Professor, and Planning Commission Chair. His projects have won national, state, and local awards for sustainability and public involvement. He is a contributing author for the ASCE book *Engineering for Sustainable Communities*.

Christine D'Onofrio is the Director of Poverty Research at the Center for Economic Opportunity (CEO). Her work includes developing uses of CEO's alternative poverty measure for New York City and other metrics as tools for policy development in the areas of opportunity and inequality. She holds a PhD in Economics from the New School for Social Research.

Jennifer C. Duthie is an Intelligent Transportation Systems engineer for the City of Austin and manages the city's traffic management center. Dr. Duthie received a PhD and MS in transportation engineering from UT Austin, and a BS in civil engineering with a minor in operations research from Cornell University. She is a licensed professional engineer in the state of Texas.

Robert W. Edmiston is a Senior User Experience Researcher with over 25 years of experience, currently applying user-centered engineering methods within the field of transportation. Robert was one of the founding members of Seattle Neighborhood Greenways, a citizen advocacy organization that encourages the City of Seattle to empower people to get around by walking and biking. After four years of volunteering, he worked for Seattle Neighborhood Greenways for one year as the Staff User Experience Engineer, defining and evaluating the comfort of Seattle's Neighborhood Greenway network.

Ahmed M. El-Geneidy is an Associate Professor at the School of Urban Planning at McGill University. Dr. El-Geneidy received a PhD from Portland State University and an MArch and BSc from Alexandria University in Egypt. His research interests include: measurements of accessibility, land-use and transportation planning, applications of GIS in land-use and transportation planning, public transit planning and operations, intelligent transportation systems, non-motorized travel modes and behavior, and the effects of transportation infrastructure on travel behavior and land value.

Ardeshir Faghri is the Director of the Delaware Center for Transportation at the University of Delaware and Professor in the Department of Civil and Environmental Engineering. He studied at the University of Washington before receiving a PhD from the University of Virginia. His research interests include transportation systems engineering, computer methods in transportation and traffic engineering, intelligent transportation systems, and transportation in developing countries.

Steven Fleming is the founder and director of Cycle Space International, an agency researching, defining, and championing bicycle centric development. He holds and has held academic positions at the Universities of Newcastle, Canberra, and Tasmania in Australia, and Harvard and Columbia in the U.S. Scholarly publications include *Cycle Space, Architecture and Urban Design in the Age of the Bicycle* (NAi010, 2012) and *Velotopia: The Production of CycleSpace* (NAi010, 2017). Exhibitors of his design work include the city of Amsterdam and the National Museum of Australia, with press coverage from *The Guardian, The Atlantic, FastCompany, ArchDaily*, and others. He is currently a conjoint Associate Professor with the Tom Farrell Institute for the Environment, a partner in Hoi!Oslo and a brand ambassador for Shimano. Dr. Fleming received his PhD from the University of Newcastle.

Mônica A. Haddad is an Associate Professor of Urban and Regional Planning at Iowa State University. She is the Director of the Graduate Certificate in Geographic Information Systems. She received a PhD and an MUP in Urban and Regional Planning from the University of Illinois at Urbana-Champaign.

Benedict Han is a planning professional based out of the City of Seattle. He graduated from the University of California, Berkeley with an MCP and a focus on community engagement in urban design, and a BA in

Community, Environment, and Planning from the University of Washington, Seattle. Previously he worked as a transportation planner for transit agencies throughout the Puget Sound region.

Justin Healy is the owner of Real Geographics, in Bend, Oregon. He has over 17 years of experience with Geographic Information Systems in addition to expertise in urban and transportation planning, economic analysis, and environmental science.

Yuwen Hou is a transportation planner at Arcadis US. She received her master of community and regional planning degree and certificate of Geographic Information Systems from Iowa State University.

Katie A. Kam is a Project Engineer in private civil engineering practice, working on public works, transportation, and land development projects. During and after she earned her PhD and MS in Civil Engineering, Dr. Kam worked at the University of Texas at Austin's Center for Transportation Research. Prior to that, she earned a BS in chemistry from UT Austin and a BA in biology from Texas A&M, using those degrees to teach high school science. After earning a MS in Community and Regional Planning from UT, she worked as a city planner with the City of Austin, Texas for seven years.

Hamza Khan is a traffic analyst at Kimley-Horn since July of 2014. He graduated from The University of Texas at Austin, receiving a BS in Civil Engineering. While at the university, he was an undergraduate researcher at the Center for Transportation Research. At Kimley-Horn, Mr. Khan has worked on a wide array of projects including corridor analysis of I-35 in Austin, CityMAP in Dallas, modeling roundabouts, lighting designs, Intelligent Transportation Systems, signal designs, TIAs, and more. His main interests are in traffic modeling, AV/CV, and context sensitive solutions. He is an avid cyclist who promotes biking and tries to use his bicycle for commuting whenever possible.

John Krampner is a researcher specializing in poverty measurement for CEO in the New York City Mayor's Office of Operations. He has an MS in Policy Analysis and Management from the New School's Milano School of International Affairs, Management and Urban Policy.

Brian H.Y. Lee is a Senior Planner at Puget Sound Regional Council in Seattle, Washington. Prior to that he was an Assistant Professor at the University of Vermont, Burlington. Dr. Lee has a PhD in Urban Design and Planning from the University of Washington, Seattle, and an MS from Northwestern University and a BASc from the University of British Columbia, both in civil engineering.

Mark Levitan is an Adjunct Professor at Hunter College's Department of Urban Policy and Planning. Until 2014, Dr. Levitan served as Director of Poverty Research at the Center for Economic Opportunity (CEO). He led the Center's work to develop a new measure of poverty for the City. Before coming to CEO, Dr. Levitan was a Senior Policy Analyst at the Community Service Society of New York where he authored studies on poverty, joblessness, and the low-wage labor market. He has also held positions at Queens College, the New York State Department of Economic Development, the Amalgamated Clothing and Textile Workers, and the United Auto Workers. Dr. Levitan received his PhD in Economics from the New School for Social Research.

Mingxin Li is a Scientist in the Department of Civil and Environmental Engineering at the University of Delaware. He received his BS in Rail Operations Management/Logistics from Beijing Jiaotong University, his MS in Civil Engineering from Oregon State University, and his PhD in Civil Engineering from the University of Utah. His research interests include transportation operational planning, rail/transit planning and management, traffic operations and analysis. His current research interests focus on travel demand modeling and learning techniques in Intelligent Transportation Systems.

Joel L. Meyer is the Pedestrian Coordinator for the City of Austin Transportation Department. He received a MS in Community and Regional Planning from the University of Texas at Austin in 2013 and a BA in History from the University of Kansas in 2009.

Daniel Scheer is a Data Scientist with SparkBeyond. He has a PhD in Economics from the New School for Social Research and has worked in applied statistical modeling positions in non-profit, government, and private sector settings.

Todd Seidel is the Director of Homeless Programs Reporting and Analytics for the Office of Planning and Performance Management at the New York City Department of Social Services (NYC DSS). Prior to his time at NYC DSS, he worked as a Research Associate at the City of New York's Center for Economic Opportunity (CEO) where he helped develop the CEO Poverty Measure. He received his MUP from Hunter College, City University of New York.

Arthur Slabosky is a Transportation Engineer in the Transportation Planning Bureau of the Michigan Department of Transportation (MDOT). He has experience in many aspects of traffic engineering, mostly as a supervisor with the Safety Programs Unit in the Traffic and Safety Division of MDOT. Mr. Slabosky is a member of the League of Michigan Bicyclists and the Tri-County Bicycle Association (Lansing, Michigan) and a member of the Association of Pedestrian and Bicycle Professionals. He has also been a member for six years, including two as chair, of the advisory citizens' Transportation Commission of the City of East Lansing. Mr. Slabosky has been a three-season bike commuter since 2001. He is a licensed professional engineer in the state of Michigan.

Cathy Tuttle is the Founding Executive Director of Seattle Neighborhood Greenways where she coordinates citywide strategic planning, branding, events, social media, advocacy, mapping, and outreach for 19 local greenway groups. She received her PhD in Urban Design and Planning from the University of Washington.

Dea van Lierop is a PhD student at the School of Urban Planning at McGill University where she also received her MUP. Her research interests include: public transit satisfaction and loyalty, land use and transportation planning, pedestrian and cyclist safety, and innovations in transit-oriented development.

Preface

This book emerged from the Bicycle Urbanism Symposium held at the University of Washington in Seattle in June 2013. Both the symposium and the book ask the question "How do we design cities for bicycles?" Over 200 participants joined the symposium from around the world, coming from Australia, Canada, China, Denmark, Hong Kong, Ireland, the Netherlands, New Zealand, the United Kingdom, and the United States. The symposium brought together practitioners, academics, policy makers, and advocates with diverse backgrounds including urban design, planning, transportation, engineering, landscape architecture, public policy, and advocacy.

The overarching premise of the symposium was to bring together top experts and advocates with the goal of creating useful scholarship. In the symposium, participants explored the ways that cities can best encourage and accommodate bicycle travel in the future. Speakers led sessions on topics ranging from imagining the 20–30-year future of bicycle friendly cities, integrating bicycling into urban planning and design, and studying the effects of bike use on health and environment to policies for developing bicycle infrastructure and programs, best practices in bicycle facility design and implementation, advances in bicycle and gear technology, economic considerations, and implementing bicycle policies and plans. This book emerged from this exploration of new research and is built from a select number of papers solicited to cover a range of topics.

The authors in this book explore dimensions of how best to redevelop cities so that they support bicycling. They use the concept of bicycle urbanism in several different ways. First, they use the presence of bicycles in the city as a measure of the "fitness" of the built environment to perform well physically and to support human relationships and health. Second, they use the desire for a bikeable city to guide urban design and development decisions that help us move toward equitable access to transportation options—a crucial component of personal mobility—as well as to shape cities that are human scaled and connected to the human experience. Third, they look at streets as the democratic medium that they are and attempt to design our rights-of-way to help build cities that work well for all people.

Acknowledgments

Thank you to Professor Emeritus Donald Miller of the Department of Urban Design and Planning at the University of Washington, Seattle, who envisioned and brought to reality the 2013 Bicycle Urbanism Symposium. Thanks also to Affiliate Assistant Professor of the Department of Civil and Environmental Engineering Alon Bassok, who took on the early duties of editing this volume. To the authors, I thank them for their enthusiasm for this project. I want to acknowledge and thank the Johnston-Hastings family for their forward-thinking publication support for UW College of Built Environments faculty. I extend my gratitude to the editorial and production teams at Routledge and Florence Production Ltd for their support. Thank you to Kim Voros and Erin David of Alta Planning + Design and Ali Masterson, a master of urban planning student in the Department of Urban Design and Planning, for graphic and data support. A final thanks to Manish Chalana, Andy Dannenberg, Don Miller, Gloria Thomas, and Nicole Gurran, Co-editor of the Environment and Planning Series for their feedback on this project.

Advancing Bicycle Urbanism

Rachel Berney

Over the past several decades, bicycling has received renewed interest as a mode of affordable transportation and a means to improve health as well as to reduce the environmental impacts associated with transportation. Much of the discussion surrounding cycling has centered on bicycle facility design—how to best repurpose road infrastructure to accommodate bicycles. And, while part of the discussion has touched upon the culture of cycling and making it a part of everyday life, urban design and city planning consider-ations have not yet fully entered the conversation. This volume presents a departure: it is focused on the city, not just the streetscape or the infra-structure.

Rather than addressing the best ways to build roads and trails to accom-modate bicycling, this book asks how best to design and build cities for bicycles. One part of the response to that question is to focus on the quality of the built environment and how it is designed; another part is to look at how cyclists use the fabric of the city, how well it serves them, and how we can evaluate it to make it better; and yet another part is to consider how well the city supports people's mobility, how well it provides equitable access to bicycling as a safe, pleasurable, and useful transportation option.

By prioritizing a bikeable city, we allow bicycling to serve as a standard in urban-development processes; that is, we use the quantity of bikers and their cycling habits as benchmarks for the "fitness" of the built environment to serve their needs. Through this practice, we can guide and evaluate urban-development decisions and changes to the built environment to better align the city with the human scale and human experience as well as to ensure equitable access to transportation options. We can also better integrate larger-scale changes, such as land-use designations and urban planning and design projects, into the right-of-way network.

By doing this, bicycle urbanism also contributes, overall, to better-designed active transportation networks that support human-powered transportation modes and promotes better facilities for walking and rolling, among other uses. Advancing bicycle urbanism also recognizes that many people are multi-modal on a daily basis and helps to promote networks and facilities that support easy transitions from walking and biking to light rail and other modes. Bicycle urbanism supports good city design.

While this volume is about cities overall, I should of course highlight the special role of streets, one of the most important ingredients of good neighborhoods and cities. They are the most durable and adaptable urban spaces. In U.S. cities, the public rights-of-way frequently take up 20 to 30 percent of the land area—a major piece of a city. By considering bicycling to be one of the primary uses of street space, planners can more effectively build cities for everyone.

This book joins an ongoing conversation of distinct but interwoven voices on cycling in cities. Some researchers look holistically at bicycling, presenting case studies from many different places. John Pucher and Ralph Buehler's seminal *City Cycling* (2012) is an excellent example. Other books help promote bicycling by dealing with infrastructure, including exploring ideas on transforming existing transportation infrastructure to better accommodate bicycling as well as discussing best practices for bicycle infrastructure design. Representative volumes include Elly Blue's *Bikenomics: How Bicycling Can Save the Economy* (2016) and Stefan Bendiks and Aglaée Degros's *Cycle Infrastructure* (2013).

Also complementary to the discussion around building cities for bicycles are books on facility design, as exemplified by John Forester's *Bicycle Transportation: A Handbook for Cycling Transportation Engineers* (1994) and the National Association of City Transportation Officials' *Urban Guideway Bike Guide* (2014). Yet other volumes inspire us through history, cultural expression, and design futures. Peter Jordan's memoir *In the City of Bikes: The Story of the Amsterdam Cyclist* (2013) combines a personal account with the local biking history of Amsterdam. There is Carleton Reid's *Roads Were not Built for Cars: How Cyclists Were the First to Push for Good Roads and Became Pioneers of Motoring* (2015). Luis A. Vivanco's 2013 book *Reconsidering the Bicycle: An Anthropological Perspective on a New (Old) Thing* brings an anthropological viewpoint to the resurgence of bicycling in recent years. Steven Fleming's 2012 *Cycle Space: Architecture and Urban Design in the Age of the Bicycle* focuses on the relationship between architecture and bicycling.

The authors in this book hope to contribute to the conversation on improving urban biking by looking at the issue from new angles and exploring forward-looking ideas that engage the technologies available now to lead city dwellers into a more resilient and healthier future.

The state of Seattle's bicycle urbanism

Given that the University of Washington hosted the Bicycle Urbanism Symposium, it seems appropriate to use the state of Seattle's bicycle urbanism as an example for consideration of ways to make all cities more bikeable. In the 1990s, Seattle was *the* biking city, arguably the burgeoning center of bicycle urbanism in the United States. At that time it boasted a bike-commute rate of 1.5 percent, versus just 1 percent in Portland, Oregon (Pucher 2013). In 2007 the city approved its first Bicycle Master Plan, setting in motion a ten-year process of studying Seattle's cycling infrastructure and setting out plans for improvements. By 2013, when the symposium was held, however, other cities had surpassed Seattle in terms of numbers of people biking as well as amount of infrastructure built. Portland's bike-commute rate was up to 6.9 percent, while Seattle's had only risen to 3.7 percent (Pucher 2013). And big city mayors from around the country, such as Chicago's Rahm Emanuel, were eyeing Pacific Northwest cities with the hope of attracting their tech workers back east—with better bicycling facilities! Emanuel said, "I want [PNW cities] to be envious because I expect to not only take all their bikers, but I'll take their jobs that come with this, all the economic growth that comes with this and the opportunities that come with this" (Schlabowske 2013). Seattle's current Bicycle Master Plan, which reaches completion in 2017, acknowledges that "to compete and attract talent, Seattle has to be a better biking city" (Seattle DOT 2014, 1).

Seattle has some major challenges when it comes to becoming a truly bikeable city. Its biking environment can be dangerous, especially in the frequently rainy conditions and during the winter, when the area experiences low levels of morning and evening light. The city is also hilly and punctuated with water bodies; these geographical constraints limit where people can ride (see Figure 0.1).

Some of Seattle's bicycle infrastructure is not well designed and ought to be better. The city is overly dependent on sharrows (streets that have bike icons and arrows painted on the pavement to remind drivers and bicyclists to share the road) and unprotected bicycle lanes. And, in comparison to other biking cities, it has relatively few women cycling. Other cities, such as Vancouver, British Columbia, and Portland, have inclusive bicycle programs, some targeted specifically to female riders. This is important because women are the "indicator species" of a cycling community; that is, "the proportion of cycling trips made by females is an important indicator to measure how safe cycling conditions are perceived to be" (Vancouver Metro Translink 2011, 27).

While in Seattle for the Bicycle Urbanism Symposium, John Pucher, the keynote speaker, was interviewed by the *Seattle Times* on the state of bicycling in the city. Pucher said he found Seattle unpleasant, even dangerous to cycle in, noting that Second Avenue downtown is "as bad as a major avenue [in] Manhattan . . . I think it's maybe even worse, because I think here, there's more left and right turns, there's more doors that are being opened, more cars that are trying to park." He went so far as to call his trip down Second Avenue "death defying." Citing design faults with the bike lane—encroachments by cars crossing the lane to enter or leave parking spaces, car doors protruding into the bike lane, and vehicles making left turns across the lane—Pucher said Second Avenue is "an egregious example of a

Figure 0.1 Seattle's geography and topography reduce options for routes and
create pinch points. Bicyclists from neighborhoods, represented by
circles proportional to their population, are channeled into the
downtown through a limited number of routes that provide for safer
travel and account for hills, bridges, and existing canal crossings.
The lines in this image represent the convergence of bicyclists from
multiple neighborhoods and the paths bicyclists are likely to take
to reach downtown.

Source: Image created by Alta Planning + Design.

poorly designed bike lane" (in Lindblom 2013). Fifty-six bicycle accidents were recorded along Second Avenue between 2007 and 2013 (Lindblom 2013). And in 2014, the death of a young attorney, who left behind her partner and child, was especially poignant as it happened within weeks of the start of a complete design overhaul of the lane (Lindblom 2014). The overhaul transformed the Second Avenue bike lane to a protected lane, or cycle track—typically a single- or dual-way lane that is painted, signed, signalized, and buffered from traffic.

Outside of downtown, Seattle is highly reliant on sharrows, which account for 34 percent of its network (see Table 0.1). But sharrows are most appropriate for neighborhoods rather than for travel across a city. In addition, Seattle's unprotected bicycle lanes and signed routes account for 30 percent of its network. Seattle tripled its mileage of bike lanes to 78 miles between 2007 and 2013, but much of that suffers from design problems that make them risky to use, including narrow widths and potential conflicts with car doors. The amount of miles of unprotected bike lanes in Seattle did not change between 2013 and the end of 2016.

Women currently comprise only two out of ten bicycle riders in Seattle (Broache 2012). That number could and should be higher. Pucher noted that "fewer than 30 percent of Seattle bike trips are made by women. It's striking that more women bike to work in Portland (4.8 percent) than the share of men biking to work in Seattle (4.6 percent)." He also pointed out that several cities, including Portland; Vancouver; Montreal, Quebec; Ottawa, Ontario; Minneapolis; and Washington, D.C., have all surpassed Seattle in the rate of bicycling by women as well as in overall bicycling (Pucher 2013).

Although Seattle's bicycle infrastructure network has grown substantially over time, it is still fragmented. It has network gaps (route gaps of less than one quarter of a mile), corridor gaps (greater than one quarter of a mile) as well as intersection gaps (intersections that require fundamental bicycle-related improvements). The relative lack of cycle tracks is sorely felt. It is interesting to note that a motor vehicle network that looked like this would be considered "broken." Figure 0.2 depicts the 171-mile bicycle network in

Table 0.1 Type, length, and percentage share of bike lanes in the Seattle network, 2016

Type of bike lane	Length in miles	Percentage of system
Cycle track (protected lane)	14.2	5
Trail	52.6	20
Bike lane	76.7	30
Sharrow	88.2	34
Neighborhood greenway	28.0	11
Total	259.7	100

Source: 2010–2015 American Community Survey 5-year estimate and Seattle Department of Transportation, bike network data current as of December 2016.

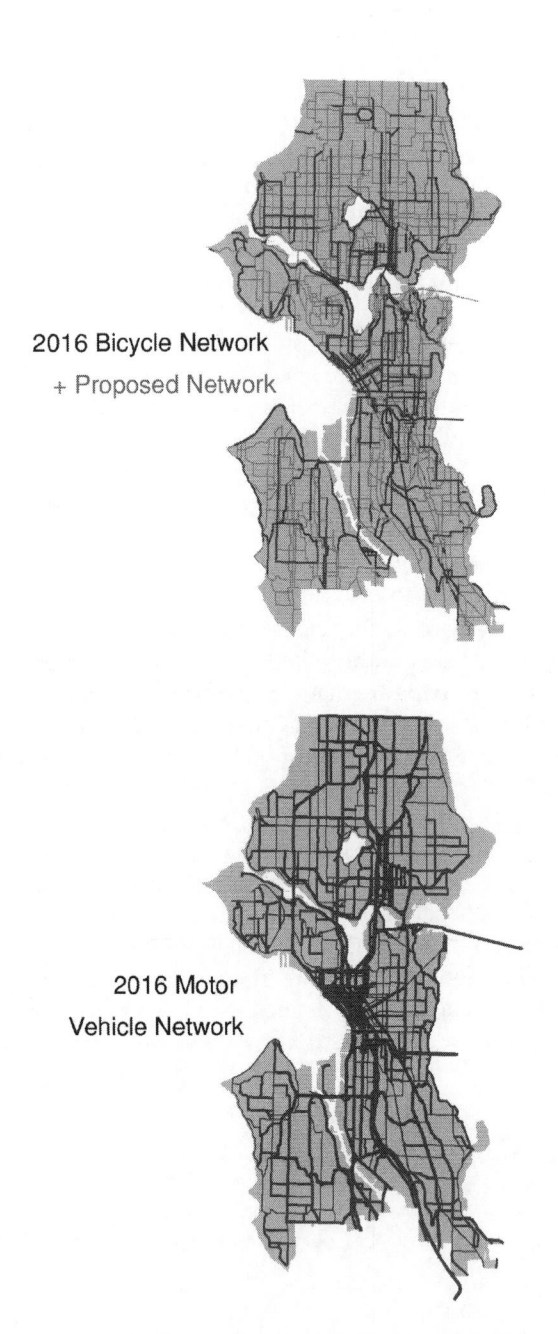

2016 Bicycle Network

+ Proposed Network

2016 Motor

Vehicle Network

Figure 0.2 This comparison of bicycle and vehicle networks in existence in Seattle in December 2016 shows that bicycle riders have about one-third the amount of linear roadway of car drivers. The dedicated bicycle infrastructure shown here is 171 miles long (planned infrastructure is shown in gray), and the motor vehicle roadway is 505 miles long.

Source: Image created by Alta Planning + Design.

Seattle, including cycle tracks, separated trails, and bike lanes, and the three-times-larger motor vehicle network, including all city arterials, implemented as of December 31, 2016.

Connectivity is strongly related to equity. Some people have great access to bicycle infrastructure in Seattle; others do not (Table 0.2). Table 0.2 shows that only 14.3 percent of Seattleites live within a 1/4 mile of a protected bicycle lane. This is mirrored by an inequitable distribution of collective bicycle resources that occurs in many communities in the U.S. (Zavestoski and Agyeman 2015). Making Seattle a more bikeable city means creating equitable opportunities for all neighborhoods to become bicycle friendly within their borders as well as in connection with longer, citywide routes that can be used for commuting. By increasing equity, the city can also increase opportunities for achieving public health benefits through bicycling. The health benefits of bicycling are well documented (Sallis, Millstein, and Carlson 2011; Garrard, Rissel, and Bauman 2012).

Another key concern for bicycling is "stress" (see Figure 0.3). Several things contribute to real or perceived difficulties when bicycling, including weather, topography, geography, and real or perceived distance and safety. Safety can be affected by factors such as road conditions, traffic volume and speed, and the level of bicycle infrastructure, as well as other built environment elements, including the number of intersections, sight lines, and place quality. The perceived distance a cyclist travels may differ from the actual difference based on the comfort of the facility. A trail or cycle track is low stress and will make a trip feel shorter than the actual distance traveled, while a shared roadway with high traffic volumes increases the perceived distance traveled. Given that 41 percent of bicycle trips in U.S. cities are less than two or three miles long (Seattle DOT 2014, 3), there is a significant room for improvement in the number of people biking if facilities were to be improved to reduce stress, but for now, for many people, even these short trips are considered "too long."

Seattle's vision is to provide the conditions that make cycling "a comfortable and integral part of daily life . . . for people of all ages and abilities" (Seattle DOT 2014, 1). The city is wisely focused on the biggest demographic of

Table 0.2 Distance of Seattle residents from protected bicycle lanes, 2016

Distance in miles from protected bicycle lanes	Number of residents	Percentage of total population
0.25	93,997	14.3
0.5	212,201	32.3
1	379,893	57.8
2	571,429	86.9
3	646,605	98.4

Source: 2010–2015 American Community Survey 5-year estimate and Seattle Department of Transportation, bike network data current as of December 2016.

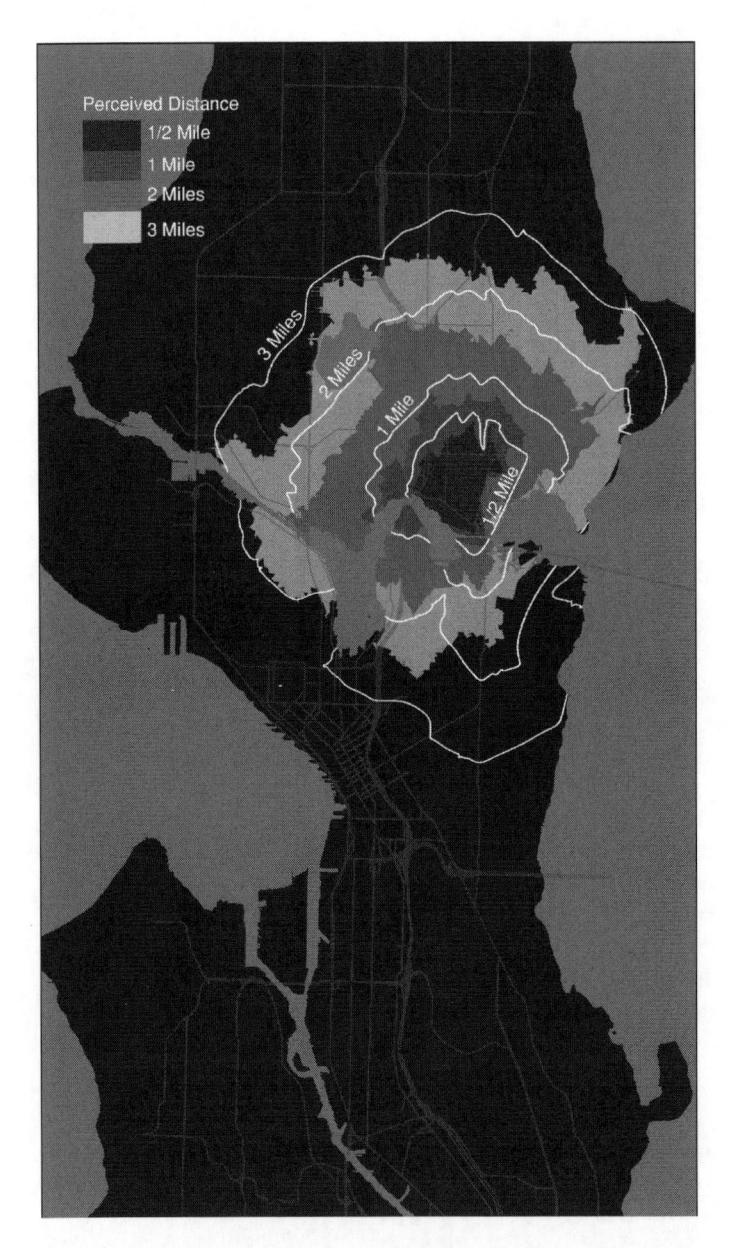

Figure 0.3 This "stress" map shows travel distance to the University of Washington. The isolines represent the actual network distances along Seattle's roadways and trails, while the shaded areas represent the perceived distances. Surrounding the University of Washington campus, high-volume roadways with minimal accommodation for bicycles lead to a significant difference between the actual network distance and the perceived distance.

Source: Image created by Alta Planning + Design.

potential new riders—the approximately 60 percent of "interested but concerned" bicyclists in the general population (Geller 2006)—and on the most likely kind of trips—short trips, or those of less than three miles. Also, the city is home to a number of local government entities and advocacy groups that are highly engaged in making Seattle a hub of bicycle urbanism. These include the Seattle Neighborhood Greenways organization, the Cascade Bicycle Club, and the Seattle Department of Transportation.

Seattle already has some established infrastructure. The broader area is home to the popular multi-use Burke-Gilman Trail, a 27-mile-long recreational trail built within a former railroad right-of-way (see Figure 0.4a). New and significant infrastructure projects are also coming online, including the protected bicycle lane on Broadway—a separated and signalized two-way cycle track in the Capitol Hill neighborhood that opened in 2014 along with the First Hill streetcar line (see Figure 0.4b)—and additional bicycle infrastructure tied to the expanding light rail system (see Figure 0.4c). A bike trail across the new SR 520 bridge is scheduled to open in later 2017.

Ideas for making the bikeable city a reality now

The chapters in this volume highlight a multitude of inventive ways to make cities more bikeable, bringing everything from technology to advocacy into the equation. Steven Fleming seeks to cultivate bicycle urbanism by looking to the aged and underutilized, but well-networked, spaces of

Figure 0.4a Seattle's Burke-Gilman Trail in the Hawthorne Hills neighborhood.

Figure 0.4b Protected bicycle lanes on Broadway in the Capitol Hill neighborhood.

Figure 0.4c Bicycle ramp connecting the upper and lower levels of the light rail station at the University of Washington.

industrial activity. By co-opting these spaces, cyclists create a new world for themselves, one that can be enhanced with new development. Robert W. Edmiston and Cathy Tuttle address the state of active transportation in Seattle. Edmiston's chapter focuses on education and advocacy work through the creation of "personas" used to engage "interested but concerned" bicyclists (Geller 2006). Tuttle's comparison of intersections in two Seattle neighborhoods discusses how signal timing for intersection crossings was tipped toward motor vehicles in a lower-income community of color, while the timing favored pedestrians—and by extension bicyclists—in a more affluent white neighborhood.

Benedict Han examines how Instagram can be used in conjunction with user surveys to evaluate and design a multi-use urban trail in Berkeley, California, while Todd Seidel, Mark Levitan, Christine D'Onofrio, John Krampner, and Daniel Scheer trace the increase in bike commuting in New York City and explore demographic characteristics and incomes of bicycle commuters below 150 percent of the poverty threshold.

Derek Chisholm and Justin Healy explore the proliferation of electric bikes (e-bikes) in urban life and their gradual adoption as commuter vehicles. This new "middle mode" has unique urban design challenges and opportunities; their chapter lays out how urban planning and design can respond to and embrace them. Arthur Slabosky presents an argument on how achieving walkable communities is dependent upon moving beyond the discussion of using speed limits as a way to control vehicles. He argues that this move, which appears counterintuitive to some, is at the heart of concerns about pedestrian- and bicycle-friendly communities.

Mingxin Li and Ardeshir Faghri propose and explore a comprehensive methodology for conducting cost–benefit analyses to reduce the knowledge gap about the level and effectiveness of health gains and consumer cost savings of added cycling facilities. Dea van Lierop, Brian H. Y. Lee, and Ahmed M. El-Geneidy explore means to reduce bicycle theft in Montreal and related vandalism that discourages bike usage based on the willingness of riders to help pay for secure bike parking. Yuwen Hou and Mônica A. Haddad explore options for a bike-share program for Iowa State University and the City of Ames, Iowa. The authors employ a mixed methodology that brings together a geographic information system (GIS) and a community survey to identify locations for possible bike stations.

Joel L. Meyer and Jennifer C. Duthie explore how planners' need to have data regarding how and where bicyclists are riding on the local street network can be met by smartphone-based data collection. This chapter demonstrates a number of ways in which GPS route data can be used to answer planning questions. Katie A. Kam, Joel L. Meyer, Jennifer C. Duthie, and Hamza Khan continue the discussion of smartphone-based data collection, turning to the need to correct for and enhance the accuracy of that data, especially as it is mapped to network files. Compared to the inherent errors in tracking movement with GPS, this chapter describes the success and accuracy achieved when using a GIS-based method to determine the roadways and off-road paths taken by cyclists.

*   *   *

The chapters in this book respond to the urgent need for bicycle urbanism *now* through proposing changes in urban built environments, including rethinking evidence-based design and planning for locating and developing infrastructure and facilities, and the ways that planning happens. While the main theme of the book and many of the chapters focus on bicycling, making cities more bikeable means making cities more friendly to active transportation networks, generally, and to multi-modal support for commuting and recreating. Bicycle urbanism helps to establish a benchmark for city form and function in which one key litmus test is the ability of people to make their way to where they need and want to go without a car.

This book presents new methods of gathering, processing, and disseminating qualitative and quantitative data, and, relatedly, underscores the importance of data gathering for advocacy purposes and to expand our understanding of who is walking and bicycling, including looking at the so-called invisible riders and their needs, desires, and behaviors that ought to be accommodated in our networks. In the same vein, it presents innovations in advocacy and engagement practices. Finally, this book considers new approaches to examining costs and benefits, as well as the willingness of bicyclists to pay for certain amenities.

References

Bendiks, Stefan, and Aglaée Degros. 2013. *Cycle Infrastructure*. Rotterdam, Netherlands: nai010.

Bhatt, Sanjay. 2013. "Amazon Gives a Push to Biking Downtown." *Seattle Times*, 13 August.

Blue, Elly. 2016. *Bikenomics: How Bicycling Can Save the Economy*. 2nd ed. Portland, OR: Microcosm.

Broache, Anne. 2012. "Perspectives on Seattle Women's Decisions to Bike for Transportation." Master's thesis, University of Washington.

Fleming, Steven. 2012. *Cycle Space: Architecture and Urban Design in the Age of the Bicycle*. Rotterdam, Netherlands: nai010.

Forester, John. 1994. *Bicycle Transportation: A Handbook for Cycling Transportation Engineers*. 2nd ed. Cambridge, MA: MIT Press.

Garrard, Jan, Chris Rissel, and Adrian Bauman. 2012. "Health Benefits of Cycling." In *City Cycling*, edited by John Pucher and Ralph Buehler, 31–55. Cambridge, MA: The MIT Press.

Geller, Roger. 2006. *Four Types of Cyclists*. Portland, OR: City of Portland Office of Transportation.

Lindblom, Mike. 2013. "Worse than Manhattan? Bike Expert Rattled by Ride through City." *Seattle Times*, 25 June.

Lindblom, Mike. 2014. "Cyclist Killed Days before City to Upgrade Notorious Bike Lane." *Seattle Times*, 30 August.

National Association of City Transportation Officials. 2014. *Urban Guideway Bike Guide*. 2nd ed. Washington, D.C.: Island Press.

Pucher, John. 2013. "Guest: Building a Bicycling Renaissance in Seattle." *Seattle Times*, 13 July.

Pucher, John, and Ralph Buehler. 2012. *City Cycling*. Cambridge, MA: MIT Press.

Sallis, James F., Rachel A. Millstein, and Jordan A. Carlson. 2011. "Community Design for Physical Activity." In *Making Healthy Places: Designing and Building for Health, Well-being, and Sustainability*, edited by Andrew L. Dannenberg, Howard Frumpkin, and Richard J. Jackson, 33–49. Washington, D.C.: Island Press.

Schlabowske, Dave. 2013. "Will Chicago's Cycletracks Steal High-Tech Jobs from Portland and Seattle?" The Bicycle Blog of Wisconsin, 18 February. http://bfw.org/2013/02/18/will-chicagos-cycletracks-steal-high-tech-jobs-from-portland-and-seattle/.

Seattle DOT (Department of Transportation). 2014. *Seattle Bicycle Master Plan*. Seattle: Seattle DOT. www.seattle.gov/transportation/docs/bmp/apr14/SBMP_21March_FINAL_full%20doc.pdf.

Vancouver Metro Translink. 2011. *Cycling for Everyone: A Regional Cycling Strategy for Metro Vancouver*. Vancouver, BC: Vancouver Metro Translink.

Vivanco, Luis A. 2013. *Reconsidering the Bicycle: An Anthropological Perspective on a New (Old) Thing*. New York: Routledge.

Zavestoski, Stephen and Julian Agyemon. 2015. "Complete Streets: What's Missing." In *Incomplete Streets: Processes, Practices, and Possibilities*, edited by Stephen Zavestoski and Julian Agyemon, 1–13. New York: Routledge.

1 Bike paths to nowhere
Bicycle infrastructure that ignores the street network

Steven Fleming

There are rare neighborhoods, streets, and venues, where homosexual couples can walk hand-in-hand and not feel self-conscious. Outside those safe places, it is not so easy to display same-sex affection without feeling threatened, or provocative. From the architectural writer Aaron Betsky, we take the term "queer space" (Betsky 1997) to describe the worlds homosexuals physically construct around themselves using clothing, objects, architecture, and urban districts, in order to attain fulfillment and safety.

My book *Cycle Space* (Fleming 2012) took a loan from Betsky's idea. In that book the term "cycle space" is used to describe the worlds cyclists build around themselves with clothing, accoutrements, sometimes architecture, and invariably safe routes through cities. Cognitively and actually, we build cycle space to attain our own sense of fulfillment and safety as cyclists. Because there is such a wide range of motivations to cycle—from fitness, to environmental awareness, to plain old frugality—cycle spaces are plural, with their loci in individuals' minds.

A good example of a kind of cycle space with its locus in riders' minds is the moving space that follows a critical mass ride. Similarly, lycra wearing members of bunch rides can turn the busiest of roads into places of fulfillment and safety. However, it is the quotidian problem of commuting alone at night in the rain that sees most regular cyclists retreating to non-vehicular routes wherever we can.

There is a danger here that habituation to non-vehicular routes can give us a very different image of our cities to those held in the minds of our neighbors who use the road network. In *The Image of the City*, Kevin Lynch argued for mental images held in common. Developing an unusual cognitive map by retreating to non-vehicular routes could cause an individual to lose their ability to "operate successfully within his environment and [. . .] cooperate with his fellows," (Lynch 1960, 46).

Ideally all cities would weld safe and fulfilling bicycling networks to the street network and give cyclists the same image of the city as the one shared by drivers, pedestrians, and frequent bus users. Unfortunately, networks of that sort are mostly only being built in cities so crippled by congestion and shortages of car parking that most voters have abandoned using a car for most trips. These are cities such as New York, Paris, and London. Even here, change is happening slowly. In thousands of other much smaller cities, change is happening at glacial pace. Voters in small cities remain satisfied overall with car dependence, so naturally elect politicians who promise more car lanes, not bike lanes. In small cities, cyclists look doomed to be bullied from streets to the margins, for as long as there are fuel sources.

Working through the five elements that, according to Lynch, define our images of our cities, we can generalize that "paths" in the minds of most cyclists would incorporate a few easements corresponding to former bulk haulage (freight) routes; that "edges" for cyclists are less likely to be shop-lined than lined with blank walls, graffiti, and wire fences; that cyclists' "districts" could include derelict zones that neither industry nor mainstream car-dependent society are actively contesting; that bike riders' "nodes" will have no markers put there by governments; and that something like a culvert section under a highway is more likely to be a "landmark" for cyclists than any civic monument or sign of commerce.

Ideologically, we can object to cyclists being banished from the street network. Urban designers from Jane Jacobs to Jan Gehl have campaigned to reclaim the street for vulnerable modes. So magnetic is the street as a topic that even when a research article purports to be about linear parks, it will often swing around to the topic of streets and how they too could be networks of greenways.

That trend begins with Anthony Walmsley's attack on writers before him, Little 1990 and Lynch 1981, for ignoring the notion of "recovering the most immediate public open space of all—the principal streets of the city—as tree-lined routes" (Walmsley 1995, 82). In like fashion, Karl Kullmann's 2013 paper about linear voids in post-industrial cities having second lives as circulation systems leads into a lengthy discussion of car-accessible streets in Portland, Oregon that have recently had a change of name from "bike boulevards" to "greenways" (Kullmann 2013), despite them being nothing more than traffic-calmed roads.

Writing from the critical stance of bicycle advocacy, I naturally support the grand project of reclaiming streets from automobiles and recognize the environmental, epidemiological, and economic benefits of non-car-focused streets. At the risk of seeming churlish though, I need to draw attention to one disadvantage. This cause that unites us—reclaiming the street from the automobile—stands in the way of pragmatic discussions about routes through our cities that could be even more helpful to cycling.

Streets provide access to building stock that was not created with bike transport in mind. By contrast, post-industrial voids through our cities could unlock brownfield redevelopment sites for genuinely bike-friendly buildings. In this context, a "bike-friendly building" would be purpose-designed to encourage cycling, the way a garage-fronted house in the suburbs is designed to encourage driving.

Those in the homosexual community have shown how they can build queer space in places with names such as the Meat Packing District, or lately Hells Kitchen—names that smack of marginalization. If cyclists took inspiration from queers, then those of us living in cities where cycling is marginalized would want to build Cycle Space on land of unique importance to us. Because derelict industrial sites are typically intersected by bulk haulage easements also dating from the industrial era, logic suggests we treat entire networks of vacant brownfields and easements as cycling frontiers (Figure 1.1).

Figure 1.1 Detroit figure-ground. The quintessential post-industrial city, Detroit has a vast network of linear voids not being used for motorized transport potentially unlocking brownfields and blighted areas for bicycle-oriented redevelopment.

Source: Steven Fleming and Ben Thorp, cycle-space.com.

It has been shown in past eras that avant-garde architecture can help the public envision the potential of frontier lands they would otherwise never consider as places to live. Construction of Le Corbusier's Villa Savoye commenced in 1928 in what, at the time, must have seemed an unimaginably remote reach of Paris. Yet the villa's design helped people imagine a new way of life, when the freeways of the future would direct them straight into their houses. Looking at the Villa Savoye they could have imagined the house of the future incorporating a U-turning bay within its structural volume, especially for cars. Le Corbusier's vision helped whet peoples' appetites for the French *Autoroute* (highway) system.

At the same time in America, Buckminster Fuller was prototyping his Dymaxion House, similarly suited to suburban locations, not in-fill sites in the city. Like the Villa Savoye, the Dymaxion House was raised off the ground for cars to park under. Yet it would be another three decades before America would start in earnest on its national freeway construction and longer still before driving would become mainstream.

Fuller was ahead of his time, so he did not make fortunes selling Dymaxion houses. What he did do was help inspire the influential General Motors-sponsored Futurama Exhibit at the 1939 World's Fair in New York. Thousands flocked to see what would prove to be a self-fulfilling prophetic vision of the good life to come, when freeways would transport people to futuristic buildings depicted on the peri-urban frontier.

Surely architecture's persuasive power is as valid today. If we can allow ourselves to take lessons from car-generated building typologies of the interwar era, we might start conceiving buildings as destinations for bike trips, the way Corbusier and Fuller conceived them for car trips. Architects would publish images of bike-centric buildings sited on the frontier lands it is argued are inherently suited to cycling: derelict industrial sites linked by former bulk-haulage routes.

Like countless small cities worldwide, the city in which I was based from 2012 to 2015, Launceston in Tasmania, grew around a river port with rail links established to distribute goods. In the post-World War II era, attention turned to sites connected by the grey network of roads. Without the density to support non-timetabled buses or trams, and without political support for barrier protected bike infrastructure aligned to the road network (that car driving voters maintain is their turf), development of the grey network and the sites that it serves seems set to remain centered on cars. Where cycling is marginalized to such an extent its best hope of flourishing may be at the margins: on brownfields.

So let us imagine the non-vehicular easements (river banks, levees, and disused rail corridors) that radiate out from the port lands of this city have been made into bicycle routes. Connections would exist between many schools and the city's university campuses. The next task, if we were to develop a bike-focused layer of the city, would be to develop new buildings for other functions such as shopping, housing, and working. The docklands in Launceston would be the obvious site, since the bulk haulage routes we would like to see made into trails or bike highways converge at this point, and because the site has many hectares of land that are not being used.

The remainder of this chapter will present design speculations, illustrated with examples that have been designed by students of architecture at the University of Tasmania. These designs highlight the possibilities for truly bike-friendly architecture that exist on large undeveloped parts of our cities. The longer term hope is that drawings, models, and exhibitions of hyper-bike-friendly buildings on sites such as Launceston's docklands will lead planners, developers, and architects to take a fresh view of brownfields and the easements that can be used to connect them for cyclists.

Shops

How might the planning and provision of retail space differ in a bike-focused district? We can take a few cues from cities such as Groningen in the Netherlands, or Ferrara in Italy, where people use bikes to go shopping. We must remember though, that these cities were built around walking and horse drawn transportation. Shops in these cities weren't designed for customers to enter on horseback, and litter the floors with manure. If in these cities the architecture, arrangement of checkouts and aisles, and social customs had grown around cycling (rather than horses and walking) we can imagine bikes being utilized fully (Figure 1.2).

Bikes with large baskets or cargo bays might now be used instead of shopping trollies, and taken down aisles. But instead, bikes are hitched to railings outside in the so-called "bike cities" of Europe. "The silent steed" (as the bike was once called) is being punished for the bad habits of the

Figure 1.2 Author Steven Fleming shows how shopping center layouts do not accommodate the use of cargo bikes as shopping trollies.

Source: Matt Sansom.

animal it was built to replace, as though bikes would urinate on the floor of a grocery store and randomly kick at people stepping behind them.

Had European cities evolved around bikes and not walking, it is possible the shops would not have been gathered together with narrow shop fronts that speak to the speed of the walker. A key theorist on modern shopping center design, Barry Maitland, surveyed the dimensions and grain of medieval shopping streets in Europe and used these to inform his influential guide book on how shopping works (Maitland 1985). From this, developers know that retail spaces in continuous strips, ideally with anchor stores at each end forming a dumbbell, attract higher rents.

The problem from an urban design point of view is that gathering shops in select streets or malls robs other streets in a city of the passive surveillance that shopfronts provide. The opportunity in a district that is purpose-designed to let people ride between shops, and wheel bikes inside, is to separate shops while at the same time not decreasing the frequency of passing trade (Figure 1.3). There is an opportunity here to eliminate inactive zones altogether, by spreading shops evenly across new urban districts. Remembering that cyclists are not cocooned within cars but rather are insinuated in events on the street, we could be hopeful of flows of bike traffic extending passive surveillance beyond the light of the shop fronts.

Spreading shopping across a whole district instead of concentrating it in pedestrian malls, arcades, or tight streets, entails a five-fold increase in the speed of the shopper, from 2 miles per hour to 10 miles per hour. This calls for design strategies that help cyclists go faster between shops, but slow down

Figure 1.3 Kiosks in Frankfurt Airport hint at ways shops could be scattered in a bike-focused district.

Source: Steven Fleming.

where conflicts with pedestrians are likely. At least four planning strategies can be imagined to facilitate this kind of shopping at bike speed.

Two known strategies are: 1. Shepherding cyclists onto smooth paths by using rougher surfaces that cyclists don't like. Rough cobbled surfaces would act as pedestrian refuges (Figure 1.4). 2. Rounding the corners of buildings, at least at ground level, would improve sight lines (Figure 1.5).

Delineating between pedestrian paths and cycle tracks is vital along linear trails, or what Lynch would call "paths." In question here though, is a new kind of "district" with its own bike-generated urban morphology. Such a district would be crisscrossed by as many desire lines as the concourse at Grand Central Station. Overreacting to the threat of pedestrian and cyclist collisions in such a context would be like Australia's overreaction to bicyclist

Figure 1.4 Cobbles are jarring to cyclists so provide pedestrian refuge.
Source: Rob Jetson, designer.

Figure 1.5 Rounding ground-floor corners improves sight lines for slow-riding cyclists.
Source: Rob Maver, designer.

head injuries by introducing mandatory bike helmet laws. We have to trust that bikes and pedestrians would learn to mix in a non-linear district the way they mix in Italian piazzas, or as Dutch traffic engineer Hans Monderman observed different modes mixing when road markings and signage are taken away and people are left to make in-the-moment decisions, or as pedestrians and cyclists are successfully interacting in the shared space beside the ferry docks behind Amsterdam Centraal train station. Giving cyclists a sense of entitlement to ride at full pace, with a blue painted cycle track, would be confusing the needs of a *district* with the needs of a *path*.

As well as varying textures and removing blind corners, there are things we can do using contours to gently lessen conflicts with pedestrians while helping cyclists shrink space-time with speed. Looking at velodromes we can see that a bike can comfortably go in a straight line on a banked surface that would put strain on a pedestrian's ankles. We also see with velodromes how kinetic energy (speed) is converted into potential energy (height) when riders move to the top of the velodrome wall, and how that stored energy is instantly converted back into kinetic energy when riders turn and ride down.

From this comes the realization that ground planes could be raised around shops and stair or lift lobbies. If slab blocks or point blocks were raised off the ground, sites could be traversed freely at bike speed on lower levels, while milling could occur at the high level. There, textured surfaces could also be used to encourage bikes to slow down (Figure 1.6).

Figure 1.6 Ground plane conceptualized as a field of moguls to slow bikes near shops and lobbies.

Source: Steven Fleming, designer, cycle-space.com.

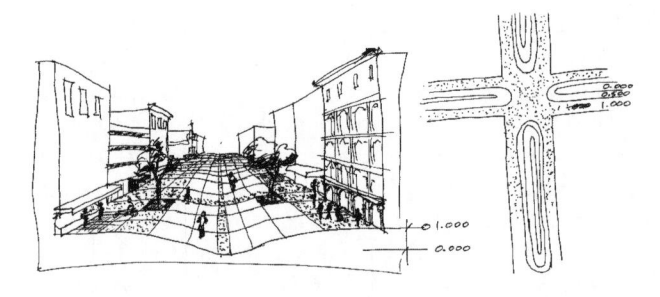

Figure 1.7 Sketch of a street scalloped to increase bicyclists' speeds as they veer
      toward center.

Source: Steven Fleming, designer, cycle-space.com.

Where the legibility of the linear street is preferred, cobbled edges could
be combined with smooth scalloped centers (Figure 1.7).

Given reservations some might have about safety, access for trikes, and
problems with ice, I must emphasize that I'm not recommending ideas such
as these be immediately put into practice. I'm recommending they be put into
visions. Avant-garde visions for new typologies and morphologies, which are
unto cycling as the Villa Savoye was unto driving, would highlight the value
of derelict industrial zones.

Housing

On brownfield redevelopment sites we can imagine housing stock that
encourages bike trips the way taps and basins in houses encourage people
to wash their hands after the toilet (Figure 1.8a–b). Life expectancy was
shorter when buildings made hygiene a chore. We can use the same argument
to say that life expectancy is shorter right now, because our buildings make
it a chore to leave home on a bike.

The architect Bill Dunster from ZEDfactory was the first to conceive
mass-housing types designed around bicycle access to every apartment. His
proposal called Velocity in London, would have seen cyclists taking a lift
to the top of a ramp system then winding their way back to their apartment,
then to the ground. Bjarke Ingels Group has been able to actually build something along similar lines, with its 8-House on brownfield redevelopment
land south of Copenhagen. Cyclists living in maisonette apartments opening
onto a ramped aerial street can ride from the uppermost levels to the ground
without encountering a stair or a lift. The building achieves the density
required for bike transport to thrive plus the views and sun access demanded
by buyers, while at the same time giving cyclists the convenience of a house
on the street. Only half of the apartments open onto the ramp through. The
rest would force cyclists into a lift with their bike or else use a storeroom
below. The convenience of just leaving home on your bike would be lost,

Figure 1.8a Coiled-row Housing. Inspired by Velocity and 8-House, my office has been exploring building types that give residents in multi-story apartment blocks the opportunity to leave home on a bike as conveniently as if they lived in row houses. Imagining a row of two story terrace houses being tilted then coiled is the easiest way to conceptualize one such type of bike centric building.

Source: Steven Fleming, designer, cycle-space.com.

Figure 1.8b Courtyard.

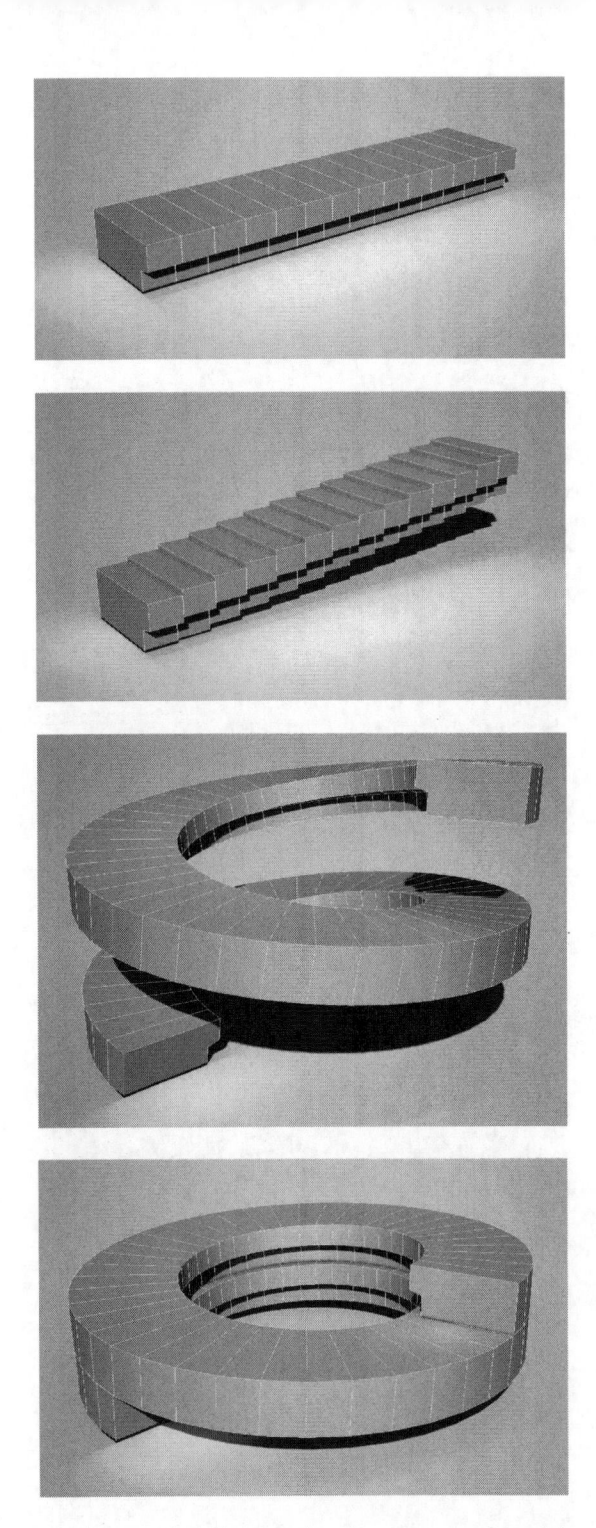

Figure 1.8c–f Coiled-row housing conceptual models.

and with it the incentive for cycling rather than driving or taking the train, neither of which provide the same health benefit as commuting by bike.

This and other types give residents a front door onto a continuous aerial street that spirals all the way to the ground (Figure 1.8c–f). Putting bike circulation on the inside radius removes the possibility of colliding forcefully against a glass balustrade, which remains a distinct risk at BIG's 8-House.

Coiling row housing is a bold, and some might say a blunt proposition (Figure 1.9a–d). Depending on our cultural backgrounds the idea will evoke Brutalist excesses such as Park Hill in Sheffield, or Alexandra Road Housing in Camden, or we might see it growing out of chic European design trends. If seen in the light of the 8-House, the aerial street has positive connotations. If populated by frequent bike users with good access to greenways, and not built solely to relocate poor people from slums, buildings organized around ramped aerial streets could deliver on the original promise of this architectural device. They could be genuinely public, inviting, and social.

A lighter approach to the aerial street has been taken by student designer Ian Watts, who integrates it with an overpass he proposes to a highway adjoining Launceston's docklands. Watts's aerial street can be read as a trafficable roof above podium level retail space on the ground floor, that at the same time increases the exposure of glass fronted apartments that could also be used as office suites (Figure 1.10).

Building stock designed for frequent bike users needs to take account of a preference by many to rest their bikes against the first wall of their apartments inside the front door. An internal layout designed by Matt Sansom

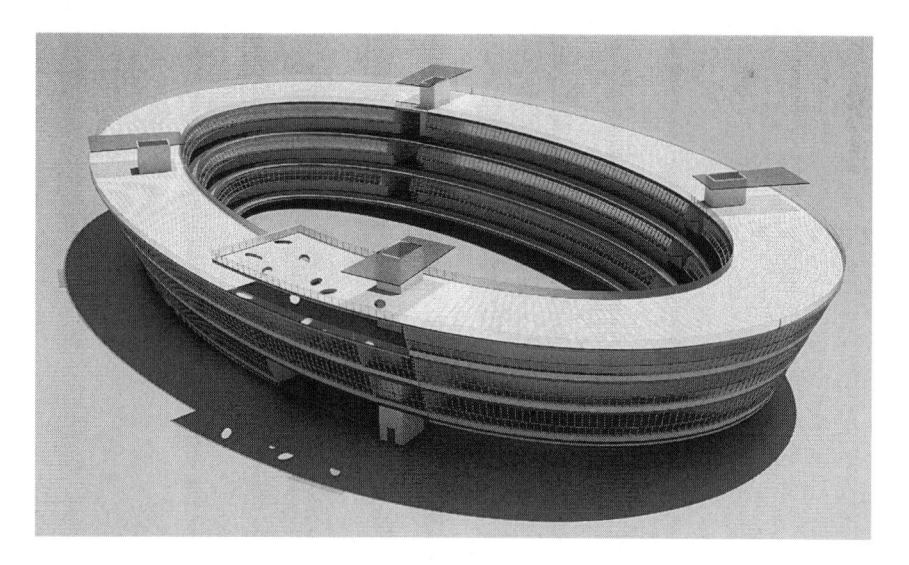

Figure 1.9a The Velohome.
Source: Abdel Soudan, designer.

provides hard-wearing surfaces to a purpose designed bike nook in that location. The length and turning radius of a long cargo bike has determined the size of the building's lift carriage and maneuvering space near the kitchen (Figure 1.11).

Figure 1.9b Velohome section.
Source: Abdel Soudan, designer.

Figure 1.9c Velohome exterior.
Source: Abdel Soudan, designer.

Figure 1.9d Velohome interior.
Source: Abdel Soudan, designer.

Figure 1.10 Aerial bicycle street married to highway overpass.
Source: Ian Watts, designer.

If nearby shops were designed so that cargo bikes could be used as shopping trolleys, parents could run errands without disturbing children asleep in these boxes that can accommodate a baby capsule (Figure 1.12, 1.13).

Figure 1.11 Apartment designed with bike nook at door and maneuvering space for a cargo bike.

Source: Matt Sansom, designer.

Figure 1.12 Shop designed so that a cargo bike can be used as a shopping trolley.

Source: Matt Sansom, designer.

Figure 1.13 Children can ride in, and sleep in, the bike that is used to do shopping.
Source: Steven Fleming. cycle-space.com

Office buildings

To understand ways that bicycle transport might impact office design, we need to look at how the type has been evolving. In advanced economies the goalposts have shifted since the days when corporate power was represented by height and architects' primary concerns were structure and the logistics of funneling thousands of workers through lifts at nine and five everyday. Now corporate tenants want to parade their *cultural* capital with green certification and signs of their concern for worker well-being. Building heights have come down, external shade structures are being added to reduce the need for air-conditioning, workers can access the outdoors by opening windows or even by stepping outside, and companies such as Google are inspiring others to treat the office more like a hangout with foosball and espresso. A number of employers in Seattle have improved cycling amenities with the specific goal of achieving gold or platinum certification from BizCycle, a body established by the Cascade Bicycle Club.

Given these trends, how might a new office building be designed in a bike-focused district? The "ride-around office building" by Tim Stocklosa looks at the extreme proposition of every worker cycling to work, parking at their workstation so their panniers can be used like extra drawers, and using their bikes to make errands within the office (Figure 1.14a, b). Clusters of workspaces are organized around a central bike ramp arranged in a continuous spiral from the secure lobby all the way to the roof, via skygardens and social spaces on every corner.

A student project by Jamie Jimenez kills two birds with one stone, by putting a sun shading bike ramp across the Western side of the lower six levels (Figure 1.15). The ramp becomes an open air breakout space, alternative access across six levels for cyclists, and most importantly shades a type of building where the typical struggle is cooling, not heating.

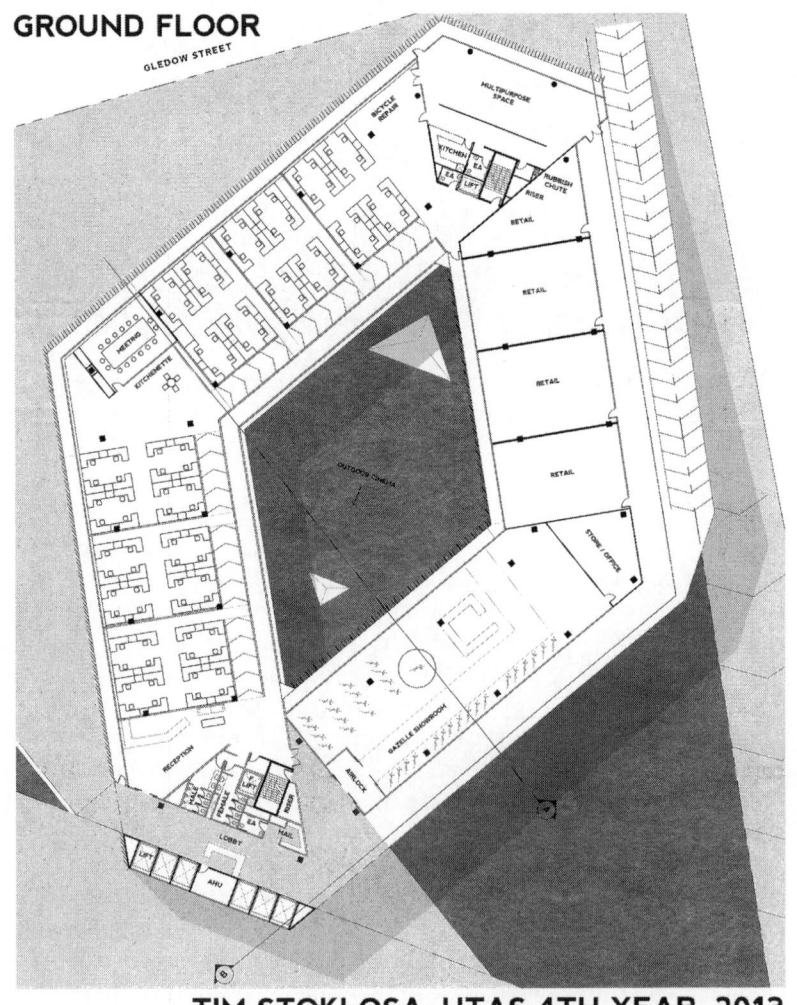

GROUND FLOOR

TIM STOKLOSA, UTAS 4TH YEAR, 2013

Figure 1.14a The ride-around office building, floor plan.

Source: Tim Stocklosa, designer.

Figure 1.14b The ride-around office building, interior.
Source: Tim Stocklosa, designer.

Figure 1.15 Office with sun-shading bicycle ramp.
Source: Jamie Jimenez, designer.

Conclusion

What does it mean to be planning and building for bikes in the post-industrial era? While most researchers and advocates will be focused on catching up with the Dutch and their bike-friendly approach to engineering the street, there is also work to be done that ignores the street network. Any city with an industrial past has enough rights-of-way and underutilized brownfields to begin sketching out a plan for an alternative cycle space city. It will have building typologies and urban morphologies that are yet to be conceived, because no city before now has made bicycle transport axiomatic to land subdivision, development controls, and architecture.

In the 1970s we could not have spoken about greenways (or linear parks) unlocking real estate development potential. Studies at that time showed that properties lost value if they were too close to parks, because parks were notorious lairs after dark (Hammer, Coughlin, and Horn 1974; Weichert and Zerbst 1973). A more recent study found the Barton Creek Greenbelt in Austin, Texas increased neighboring property values by up to 20.2 percent (Nicholls and Crompton 2005). Falling crime rates loosen some of the strictures that have bound planners and urban designers in previous decades. Conditions exist for a new paradigm in urban design, predicated on bicycle transport along trails linking redeveloped industrial districts.

The world's first Bicycle Urbanism Symposium at the University of Washington, Seattle in 2013, began with a call from Professor Donald Miller for delegates to reimagine urban population growth. I hope this chapter inspires architects and planners to reimagine population growth around design strategies that put the bike first. Fitting bikes into pedestrian settings, and car centric settings, has been tried and found wanting. The bike remains an untried mobility platform for city planning that could overcome many shortfalls of walking. It is too elegant a device to be ignored as a potential generator of urban form.

Acknowledgments

Thank you to the 2013 cohort of fourth-year students of architecture at the University of Tasmania for engaging with the task of designing a new district for bikes. Thanks too to Dr. Ceridwen Owen at the University of Tasmania and Dr. Anne Lusk at the Harvard School of Public Health for sharing their ideas and helping to clarify mine.

References

Betsky, Aaron. 1997. *Queer Space: Architecture and Same Sex Desire*. New York: William Morrow.
Fleming, Steven. 2012. *Cycle Space: Architecture and Urban Design in the Age of the Bicycle*. Rotterdam: nai010.

Hammer, Thomas R., Robert E. Coughlin, and Edward T. Horn IV. 1974. "The Effect of a Large Urban Park on Real Estate Value." *American Institute of Planning Journal* 40 (July): 274–277.

Kullmann, Karl. 2013. "Green-Networks: Integrating Alternative Circulation Systems into Post-industrial Cities." *Journal of Urban Design* 18(1): 36–58.

Little, Charles E. 1990. *Greenways for America*. Baltimore, MD: Johns Hopkins University Press.

Lynch, Kevin. 1960. *The Image of the City*. Cambridge, MA: MIT Press.

Lynch, Kevin. 1981. *A Theory of Good City Form*. Cambridge, MA: MIT Press.

Maitland, Barry. 1985. *Shopping Malls: Planning and Design*. New York: Nichols.

Nicholls, Sarah and John L. Crompton. 2005. "The Impact of Greenways on Property Values: Evidence from Austin, Texas." *Journal of Leisure Research* 37(3): 321–341.

Walmsley, Anthony. 1995. "Greenways and the Making of Urban Form." *Landscape and Urban Planning* 33: 81–127.

Weichert, John C. and Robert H. Zerbst. 1973. "The Externalities of Neighborhood Parks: An Empirical Investigation." *Land Economics* 49: 99–105.

2 Traffic signal equity
Crossing the street to active transportation

Cathy Tuttle

Introduction

One of the most significant barriers to active transportation is simply crossing the street, particularly crossing arterials (Koonce 2012). This chapter reports on a study that tests whether traffic signals in lower-income communities favor motorized vehicles more than they do in higher-income communities. Signal timing and the difficulty of crossing the street can have negative cascading effects on community well-being and public health including: perceived safety, pedestrian injury and collisions, fewer people walking and biking, higher rates of heart disease and obesity, and economic vibrancy of small business nodes (Rutt, Dannenberg and Kochtitzky 2008). Graduate students at the University of Washington's School of Public Health and community volunteers studied three sets of street crossings. Two sets of crossings are located in the Rainier Valley, Seattle's most ethnically diverse area, and a third set acted as a control in the more homogeneous and higher-income Ballard area in northwest Seattle. Students and volunteers recorded pedestrian volume, delay data, and jaywalking as well as collected key informant and user experience accounts. This study was prompted by people who live and work in the Rainier Valley who have repeatedly emphasized in public meetings that they did not have enough time to safely cross their streets. Their concerns were validated by one of the most significant findings of this study. The crossing time based on distance in feet per second had a statistically significant difference based on location. In higher-income Ballard with transit, freight, and more than 23,800 daily vehicles, pedestrians can cross the street at 2.0 to 2.8 feet per second. In lower income and more racially diverse Rainier Valley, with comparably active vehicle volumes, pedestrians had to hurry across their streets at 3.1 to 3.9 feet per second.

Health and transportation

Imagine standing on a street corner, waiting for the light to turn green. You are holding two heavy bags of groceries. Daylight is fading. It is the end of a long week. You look up at the slight hill heading towards the six-block walk to the light rail station and feel even more tired. You watch two lanes in each direction of trucks rumble by, the wheezing buses, cars waiting in the center turning lane, the impatient motorists zipping through at the last possible second as the signal finally changes to your green "Walk." If you were a transportation engineer or planner, you'd know you were crossing a freight and transit corridor carrying over 22,000 vehicles a day. You step into the street. Because it was a long day, you notice you actually have to walk slightly uphill as you cross the street. You catch the eye of a driver nosing into the intersection and about to make a right on red and he stops. You get safely to the other side of the street and walk home. It's hard to imagine doing this walk when you go with your kindergartener to her first day of school. Imagine doing this walk when you are 20 years older (Figure 2.1, 2.2).

This chapter does not address "crossing deserts"—the long expanses of heavy and fast-moving roadway people must traverse between traffic signals that are often found in areas with high economic and cultural diversity. Traffic signals have their own unique equity issues. Environments that promote active transportation are essential to the promotion of safe and healthy communities. Being able to safely cross a street on foot or by bike is critically important to a safe environment (Figure 2.3).

Figure 2.1 People waiting for signal change at Rainier Ave. S and S Henderson St.

Figure 2.2 People waiting for signal change at Rainier Ave. S and S Henderson St.

Low-income neighborhoods can have "**crossing deserts**", in much the same way that these neighborhoods often have "food deserts" (Larsen and Gilliland 2008). The following can hamper the simple act of getting across the street by walking and biking to get to work, to the store, to pick up your child at school, or simply to cross the street to greet your neighbor:

1. Fast-moving arterials with distantly placed signals that make crossing without a signal more likely;

2. Geography at intersections across arterials, even at intersections with signals, that lack engineering features that allow for safer crossing for people, including curb bulbs, advance stop lines, "no right on red" signs;

3. Signals themselves can have timing and design qualities that favor motorized vehicles over people.

It is this final category of the equity issues surrounding signal timing and design that this chapter is about, but all aspects of "crossing deserts" are a ripe area for future research.

Figure 2.3 Crossing deserts.

In 2012, Surgeon General Regina Benjamin announced the nation's first-ever National Prevention Council Action Plan, which strives to "move our health system from one based on sickness and disease to one based on wellness and prevention" (National Prevention Council 2012, 1). The plan specifically addresses transportation and encourages the development of walkable communities, bike lanes, and other healthy transit options. Public health emphasizes the value of incorporating transit in "healthy communities." Healthy communities provide residents with access to food and services; a sense of safety and community; physical activity; and clean air. Unfortunately, the transportation infrastructure in the United States still favors automobiles, which creates fewer options for people to move around on foot or bicycle and interact with services and people in their environment. The National Prevention Council (2012) indicates that a lack of alternative transportation options may have negative consequences related to physical activity, injury and prevention, air quality, and mental health status.

The ability to safely cross the street requires sufficient time to get from one side of the street to the other. Traffic signal delay can make pedestrians wait a long time for a "Walk" signal. People may become discouraged from using the crosswalk or may cross against the light. Wang et al. (2011) found that pedestrians may ignore the "Walk" and flashing "Don't Walk" signals if they do not provide sufficient crossing time. Long wait times and crossing distances, pedestrian direction of travel, long distances between crosswalks, and large numbers of pedestrians crossing contribute to negative pedestrian experiences that may lead to jaywalking (Hubbard, Bullock, and Day 2008).

According to King County, there were 90 pedestrian fatalities and 582 serious pedestrian injuries between 2008 and 2012 (Seattle and King County Public Health 2013). The National Center for Environmental Health (2012) shows that almost three-fourths (73 percent) of pedestrian fatalities occur in an urban setting and nearly 80 percent of pedestrian fatalities occur at non-intersections—commonly the result of a vehicle collision with a jay-walker. A study by Moudon et al. (2011) examining injury severity among pedestrian–motor vehicle collisions in King County found that neighborhood environment and design was a significant factor in pedestrian safety. Higher residential densities and lower median home values were associated with a higher risk of severe injury or death (Moudon et al. 2011).

When people feel unsafe crossing their neighborhood streets they are less likely to walk or bike. Improved access to neighborhood destinations has been shown to increase walking as a mode of transportation (Sugiyama et al. 2012). A safe street is a walkable street. A walkable street is a public health benefit. Sallis et al. (2009) report that adults who live in walkable neighborhoods are less likely to be overweight or obese than those living in neighborhoods with poor walking quality. See Table 2.1 for a description of the neighborhoods studied.

Table 2.1 Neighborhoods studied

Neighborhood	Type	Intersection A	Intersection B
Columbia City	Focus	S Alaska St. & Rainier Ave. S	S Edmunds St. & Rainier Ave. S
Rainier Beach	Focus	S Henderson St. & Rainier Ave. S	51st Ave. S & Rainier Ave. S
Ballard	Comparison	24th Ave. NW & NW Market St.	22nd Ave. NW & NW Market St.

Methods

Rainier Avenue South (referred to as "Rainier") is a major freight and transit corridor in south Seattle, Washington that bisects the Columbia City and Rainier Beach neighborhoods. Rainier divides dense residential areas from major public transit access points, forcing pedestrians frequently to cross at busy intersections. Rainier also functions as a major thoroughfare connecting residents of South King County to downtown Seattle and major freeways.

Columbia City in the Rainier Valley mixes small business retail, a library, a community center, and a major athletic field in its busy core. Rainier Beach boasts a mix of several public schools, a community center, library, and retail. Due to this unique combination of vibrant neighborhood activity and the fast-paced nature of Rainier as a commuter, transit, and freight route, pedestrians in these two neighborhoods have repeatedly expressed to Seattle Neighborhood Greenways planners how difficult it is to cross the street. Northwest Market Street in Ballard, the higher-income control neighborhood, is also a freight and transit corridor that runs through a thriving commercial business district. Signals are placed frequently in the Ballard commercial corridor, including some at mid-block crossings.

In February 2013, Seattle Neighborhood Greenways asked eight University of Washington Masters of Public Health students to conduct a study on the barriers people face crossing a busy arterial in two neighborhoods where residents reported difficulty crossing the street: Columbia City and Rainier Beach (Bronnum et al. 2013). While observations focused on pedestrians, this study has relevance to people riding bicycles and other forms of active transportation.

The study teams also conducted a sample of resident intercept surveys with pedestrians and key informant interviews to identify barriers associated with crossing the street. Respondents supported the claim that signal timing is a barrier to crossing the street in the two lower-income focus neighborhoods, while identifying several additional barriers related to traffic considerations, personal safety, and the built environment. Respondents in the higher-income area did not perceive signal timing as a barrier to crossing the street. From a built environment and public health perspective, signal timing, signal delay, and intersection conditions are worthy of broader research. In the immediate term, government and community action on signals is essential to facilitate positive and sustainable change in the Seattle neighborhoods of Columbia City and Rainier Beach.

The overarching research question that informed the study design was, "What are the barriers associated with pedestrians crossing the street in Columbia City and Rainier Beach?" Students conducted two studies to further understand pedestrian experiences:

1.   Signal timing: The signal timing study was designed to test the hypothesis that traffic signals favor vehicles in the lower-income focus neighborhoods studied compared to the higher-income comparison neighborhood.
2.   Community perception: The community perception study was designed to collect and analyze local informants' perceptions of the barriers to crossing Rainier Avenue South in the Rainier Beach and Columbia City neighborhoods.
3.   Control study: A third study in March 2013 by local community volunteers used intercept questionnaires in the higher-income Ballard neighborhood.

The signal timing study was designed to test the hypothesis that traffic signals favor vehicles in lower-income neighborhoods compared to higher-income neighborhoods. To determine whether there was a statistically significant difference in traffic signal timing by neighborhood, the study team examined signal delay, the length of time pedestrians had to wait at select intersections before receiving the *walk* signal and how much time they had to cross the street.

At the start of the study, data from the Seattle Department of Transportation (2011) confirmed that NW Market Street (referred to as "Market") in Ballard was an appropriate comparison from a traffic signal and traffic volume perspective to the focus neighborhoods along Rainier. The six specific intersections used in the study were selected based on location in the neighborhood, proximity to businesses, vehicle volume, and their multiple functions serving freight, transit, vehicle, and active transportation users. See Table 2.2 for select characteristics of the neighborhoods studied.

Of the three study sites, according to Seattle Department of Transportation (2013) pedestrian injuries and fatalities in study intersections were highest between 2003 and 2013 in Columbia City at 36 fatalities, followed by 22 in Rainier Beach, and five in Ballard. Signal timing, pedestrian volume, and jaywalking incidents were observed at all study intersections. Note that only in Ballard are signals automatically activated for people walking and biking, arguably an inequitable distribution of resources (Koonce 2012).

Table 2.2 Key demographic data for the Columbia City, Rainier Beach, and Ballard neighborhoods

Characteristic	Columbia City	Rainier Beach	Ballard
Race			
White (%)	33	26	85
Non-white (%)	67	74	15
Average daily vehicles	26,200	22,000	23,800
Median household income	$47,500±$12,326	$45,956±$8,214	$72,443±$5,260

Source: Seattle Department of Transportation 2011; U.S. Census Bureau 2010.

Traffic signal observations were recorded for a total of 21.5 hours over the course of three days at different time periods (morning, afternoon, and evening). All of the study intersections were statistically different. The most notable comparisons between intersections existed in the mean crosswalk distance to total time to cross ratio; mean jaywalkers; and mean pedestrian volume (see Table 2.3, Figure 2.4, 2.5). Data were cleaned and exported to SPSS (version 18) for coding and analysis. Descriptive statistics and frequencies were conducted on key variables. A linear regression was performed to determine whether the study intersections were statistically significantly different.

Figure 2.4 Study area: Rainier Ave. S and S Alaska St.

Table 2.3 Study intersection descriptions

	Ballard		Columbia City		Rainier Beach		Total
	24th & Market	*22nd & Market*	*Alaska & Rainier*	*Edmunds & Rainier*	*Henderson & Rainier*	*51st & Rainier*	
Crosswalk							
Distance (ft.)	47.9	52.3	78.6	57.6	57.1	66.0	N/A
Mean crosswalk distance/ total cross time (ft./sec.)	2.8	2.0	3.9	3.1	3.4	3.6	3.1
Frequency of observations	61	108	46	92	72	26	402
Mean pedestrian volume	2.6±2.0	5.7±4.1	1.8±1.5	2.6±1.8	2.8±2.0	1.0±0.9	3.3±3.0
Mean jaywalkers	2.4	6.3	2.2	2.4	8.0	6.1	4.6

Figure 2.5 Study area: Rainier Ave. S and S Henderson St.

Findings

The study teams also conducted a sample of resident intercept surveys with pedestrians and key informant interviews to identify barriers associated with crossing the street. Respondents supported the claim that signal timing is a barrier to crossing the street in the two lower-income focus neighborhoods, while identifying several additional barriers related to traffic considerations, personal safety, and the built environment. Respondents in the higher-income area did not perceive signal timing as a barrier to crossing the street. From a built environment and public health perspective, signal timing, signal delay, and intersection conditions are worthy of broader research. In the immediate term, government and community action on signals is essential to facilitate positive and sustainable change in the Seattle neighborhoods of Columbia City and Rainier Beach.

The concerns of Rainier Valley residents that prompted this study were validated by one of the most significant findings of this investigation. Cross time divided by distance (feet per second) has a statistically significant difference in Ballard, Columbia City, and Rainier Beach. In Ballard with transit, freight, and more than 23,800 daily vehicles, pedestrians had time to cross the street at 2.0 to 2.8 feet per second. In Rainier Valley, with comparably active vehicle volumes, pedestrians had to hurry across their streets at 3.1 to 3.9 feet per second. This finding suggests that the pace at which people must walk, on average, to successfully cross the crosswalks observed in Ballard is slower than for people crossing in the two focus neighborhoods. Additionally, the total crosswalk length or width of the arterial was longest at the four study intersections crossing Rainier (Figure 2.6).

Rainier Beach had the highest amount of jaywalking during the data collection period. Ballard had higher pedestrian volume than the two focus neighborhoods in Rainier Valley. While the mean pedestrian delay time varied by intersection and neighborhood, this variation was not statistically significant. Nearly 75 percent of all pedestrian delays across the study sites were more than 30 seconds. Walk time observation data was crosschecked with SDOT signal timing cards. The walk times provided by SDOT for the study intersections were similar to the walk times observed during the study period (Figure 2.7).

In 2009, the Manual on Uniform Traffic Control Devices presented revised Federal standards on the crossing time calculation allowing a maximum of 3.5 feet per second. SDOT began enforcing this recommendation by adjusting signal timing at intersections on a case-by-case basis (Brian Kemper, personal communication, February 13, 2013). Our study findings suggest that signal timing was addressed first in higher-income Ballard, with an advantageous 2.8 and 2.0 ft/sec for people walking and biking. Alaska and Rainier and 51st and Rainier had not been adjusted since the MUTCD 2010 revision. Signal timing at the Henderson and Rainier intersection appears to meet the new standard, but just barely.

This study also found signal delay in the lower-income communities (Bronnum et al. 2013). Pedestrian delay may also encourage jaywalking, according to Heinonen and Eck (2007). People who fail to properly utilize designated crossings are at higher risk for injury and death. These results are not generalizable to other neighborhoods or other intersections, however, as they were not randomly selected and could vary greatly from other neighborhoods in Seattle and other cities.

- A number of cars sped down 51st and made a right on red onto Rainier without stopping.

- Car ran a red through crosswalk.

- On long waits, pedestrians hit the button repeatedly and seemed visibly impatient.

Figure 2.6 Rainier Beach observations.

> • Blind pedestrians were observed several times walking against the signal. When observers questioned why, the blind pedestrians said they were walking with the audible signal meant for the opposite side of the street but focused on their side.
>
> • A mother with a toddler and a baby in a stroller was cut off by cars driving in front of her as she was trying to cross the crosswalk with the signal. This caused her not to cross during the first walk signal and to instead wait for the next signal.

Figure 2.7 Columbia City observations.

Interviews

The study team created a mixed-method, in-person interview survey tool to understand pedestrians' and key informants' perceptions of crossing Rainier. The tool was designed to determine whether signal timing is a community concern. Interview surveys were multi-phase: Phase 1 had four quantitative questions and allowed respondents to elaborate with qualitative responses; Phase 2 had two follow-up qualitative questions. The survey topics included: participant perceptions and perceived barriers to crossing Rainier.

The study team approached pedestrians (n=160) on or near Rainier in the two study neighborhoods, and a separate team spoke with Ballard pedestrians (n=42). Twelve key informant (KI) interviews were conducted in the Rainier Valley (Bronnum et al. 2013). Respondents in Columbia City and Rainier Beach reported statistically significant differences both in their primary mode of transportation and in the factors that make them feel unsafe when they cross Rainier. Respondents from Columbia City listed walking (57 percent) as their primary mode of transportation and 27 percent noted transit as their primary mode. In Rainier Beach, however, the majority of respondents (51 percent) listed transit as their primary mode of transportation and 31 percent noted walking as their primary mode. In Ballard, the majority of respondents (69 percent) listed driving as their primary transportation mode (Figure 2.8).

On a five-point Likert scale in which 1 represented "very unsafe" and 5 represented "safe," Columbia City respondents reported feeling significantly safer (mean=3.7) crossing the street than respondents in Rainier Beach (mean=3.2) when crossing Rainier. Crossing Market, Ballard residents felt the safest (mean=4.2). When respondents elaborated on what contributed to their perceived level of safety, many said it depended on location. "Location" was defined to include respondents feeling more or less safe on or in certain corners, crosswalks, and neighborhoods. Eleven respondents reported that location affects their level of safety. Four respondents mentioned that time of day influenced their safety, particularly in Rainier Beach, where pedestrians mentioned increased crime at night. Five respondents

> • "You know what I hate? When you're crossing in front of a car, a lot of times they'll speed up like they're trying to kill you or something." – Rainier Beach resident
>
> • "Speeding is equally hurtful in affecting the walkability of crossing Rainier." – Columbia City business owner

Figure 2.8 Feelings of safety.

mentioned that perceived safety was dependent on their familiarity with the area.

Signal timing was a concern for a quarter of respondents in both neighborhoods (25 percent in Columbia City; 28 percent in Rainier Beach). Interestingly in Ballard, virtually none of the respondents said signal timing was an issue (4 percent) but several mentioned nearby signals where they felt unsafe.

Similar numbers of respondents listed "other" reasons for feeling unsafe while crossing Rainier on foot (21.4 percent in Columbia City; 20 percent in Rainier Beach). Among the 83 pedestrians and 12 KIs who provided qualitative answers, "other" factors contributing to feeling unsafe included drivers not respecting traffic lights, distracted and malicious drivers, and turning cars. The most common response from pedestrians and key informants regarding their experiences crossing the street in the two focus neighborhoods involved near-miss collisions with cars (n=12); admission of or witnessing jaywalking (n=11); and actual collisions with cars (n=5). Many of the near-miss collisions involved pedestrians running out of time when crossing, turning cars failing to see or yield to pedestrians, or cars running through signals.

When asked whether they had enough time to cross at the signal, the majority of survey respondents in both focus neighborhoods (64 percent in Columbia City; 53 percent in Rainier Beach) reported they had enough time, but expressed concerns for other pedestrians in the community. A number of respondents (n=13) were concerned that older pedestrians, children, or those with limited mobility might not have enough time to cross the street.

Conclusions

The hypothesis that signal timing is a community concern was partially supported by the results. Respondents provided personal opinions regarding signal timing. Some described air pollution or making drivers angry as unintended consequences of changing the signal timing to favor pedestrians. This suggests a need for further input from additional perspectives, such as environmental health and drivers, to fully understand the advantages and disadvantages of changing signal timing. By including opportunities for respondents to provide qualitative responses, other issues not initially considered by the study team surfaced as pedestrian safety concerns. Results indicate there is a significant difference between crossing time in the higher- and lower-income neighborhoods studied. The higher-income neighborhood had significantly longer crossing times, but there was no significant difference in pedestrian delay. The findings partially support the hypothesis that signal timing in the lower-income neighborhoods studied favor vehicles. However, a direct association between neighborhood median household income and signal timing cannot be made with this limited study and results included in this report are not generalizable without further observations at more sites.

Pedestrian safety is a public health issue. Neighborhoods that support active transportation have safe and time-effective street-crossing experiences. Controlled traffic contributes to a healthy neighborhood and healthy residents. Study findings support the hypothesis that traffic signals in the lower-income focus neighborhoods were more likely to favor vehicular traffic than in the higher-income comparison neighborhood. Pedestrian delays exceeding 30 seconds are associated with non-compliance and injury; nearly 75 percent of all pedestrian delays across the focus and comparison study sites exceeded 30 seconds. People in Columbia City and Rainier Beach identified signal timing as a primary barrier to crossing Rainier. Other barriers identified include traffic, personal safety, and the built environment. Pedestrian experiences varied by neighborhood and action steps should be tailored to address the specific needs and concerns of each community.

The traffic signal functions as a support on the narrow bridge across a raging torrent of moving, jostling, distracted drivers intent first on interactions with other vehicles and then with whatever dramas are going on within their own cars. We have an obligation as planners and traffic engineers to make these narrow bridges as sturdy, safe, and reliable as possible. We can buttress the foundations of arterial crossings with curb bulbs and islands to shorten the passage of how far people need to walk or bike in the roiling water of traffic. We can focus driver attention effectively with red light cameras, distracted driver enforcement, and no right on red ordinances. Ultimately we as a society need to decide to build the best bridges possible across traffic in a way that keeps everyone focused, predictable, and alert. Well-placed, designed, and timed signals support local businesses, children walking to school, neighbors getting to know each other in our communities. While observations focused on pedestrians, this study has relevance to people riding bicycles and other forms of active transportation as well.

By adopting a public health lens, we find that there are many strategies to address pedestrian safety. Our primary recommendations for enhancing pedestrian safety in Rainier Valley include signal timing improvements; traffic infrastructure and enforcement; and community development. Collaboration with government and community partners is essential to facilitate positive and sustainable change in the Columbia City and Rainier Beach neighborhoods. More study is needed to see whether this study reflects a pattern of signal inequity in lower-income communities.

Acknowledgments

Students from the University of Washington School of Public Health Community Oriented Public Health Practice Program (COPH) were instrumental in literature searches, data collection, and analysis for this research. Peter J. House, MHA, Senior Lecturer, directed the COPH program and taught students Amber Bronnum, Clarissa Lord Brundage, Elizabeth Burpee, Nick Canavas, Jennifer Morton, Barbara Obena, Sierra Rotakhina, and Genya Shimkin, who all contributed to this community-based work. Community volunteers in the Ballard Greenways and Rainier Valley Greenways groups also participated in data gathering.

References

Bronnum, Amber, Clarissa Lord Brundage, Elizabeth Burpee, Nick Canavas, Jennifer Morton, Barbara Obena, Sierra Rotakhina, and Genya Shimkin. 2013. *Crossing Rainier Avenue: Two Studies Exploring the Pedestrian Experience in the Rainier Valley.* Seattle: University of Washington, School of Public Health. http://seattlegreenways.org/wp-content/uploads/Crossing-Rainier_Pedestrian-Experience_March-12-2013_FINAL.pdf.

Heinonen, Justin A. and John E. Eck. 2007. *Pedestrian Injuries and Fatalities.* Washington, D.C.: U.S. Department of Justice, Office of Community Oriented Policing Services. https://ric-zai-inc.com/Publications/cops-p135-pub.pdf.

Hubbard, Sarah M. L., Darcy Bullock, and Christopher M. Day. 2008. "Integration of Real-time Pedestrian Performance Measures into Existing Infrastructure of Traffic Signal System." *Transportation Research Record: Journal of the Transportation Research Board* 2080: 37–47.

Koonce, Peter. 2012. *Confessions of a Traffic Engineer: The Misuse of Level of Service.* Presentation, Equity and Health in Transportation: Metrics and Best Practices, Puget Sound Regional Council, Tacoma, WA, 28 September.

Larsen, Kristian and Jason Gilliland. 2008. "Mapping the Evolution of 'Food Deserts' in a Canadian City: Supermarket Accessibility in London, Ontario, 1961–2005." *International Journal of Health Geographics* 7: 16.

Moudon Anne V., Lin Lin, Junfeng Jiao, Philip Hurvitz, and Paula Reeves. 2011. "The Risk of Pedestrian Injury and Fatality in Collisions with Motor Vehicles, A Social Ecological Study of State Routes and City Streets in King County, Washington." *Accident, Analysis and Prevention* 43(1): 11–24.

National Center for Environmental Health. 2012. *CDC Transportation Recommendations—Brief.* Atlanta, GA: National Center for Environmental Health.

National Prevention Council. 2012. *National Prevention Council Action Plan: Implementing the National Prevention Strategy.* Washington, D.C.: U.S. Department of Health & Human Services. www.surgeongeneral.gov/priorities/prevention/2012-npc-action-plan.pdf.

Rutt, Candace, Andrew L. Dannenberg, and Christopher Kochtitzky. 2008. "Using Policy and Built Environment Interventions to Improve Public Health." *Journal of Public Health Management and Practice* 14(3): 221–223.

Sallis, James F., Brian E. Saelens, Lawrence D. Frank, Terry L. Conway, Donald J. Slymen, Kelli L. Cain, James E. Chapman, and Jacqueline Kerr. 2009. "Neighborhood Built Environment and Income: Examining Multiple Health Outcomes." *Social Science & Medicine* 68(7): 1285–1293. www.pubmedcentral.nih.gov/article render.fcgi?artid=3500640&tool=pmcentrez&r endertype=abstract.

Seattle Department of Transportation. 2011. *Traffic Flow Data and Maps*. www. seattle.gov/transportation/tfdmaps11.htm.

Seattle Department of Transportation. 2013. *Collision Diagram Report*.

Seattle and King County Public Health. 2013. *Pedestrian Safety*. www.kingcounty. gov/depts/health/violence-injury-prevention/traffic-safety/pedestrian-safety.aspx.

Sugiyama Takemi, Maike Neuhaus, Rachel Cole, Billie Giles-Corti, and Neville Owen. 2012. "Destination and Route Attributes Associated with Adults' Walking: A Review." *Medicine and Science in Sports and Exercise* 44(7): 1275–1286. www.ncbi.nlm.nih.gov/pubmed/22217568.

U.S. Census. 2010. City of Seattle. *Population Characteristics, Summary File 1*. www.seattle.gov/dpd/cms/groups/pan/@pan/documents/web_informational/dpdp 022056.pdf (Link no longer active).

Wang, Wuhong, Hongwei Guo, Ziyou Gao, and Heiner Bubb. 2011. "Individual Differences of Pedestrian Behaviour in Midblock Crosswalk and Intersection." *International Journal of Crashworthiness* 16(1): 1–9.

3 The role of personas in cycling advocacy

Robert W. Edmiston

Introduction

This chapter explains how a grassroots bicycle and pedestrian advocacy organization in Seattle, Washington, used personas as a tool to make the latest bicycle transportation research intuitively usable by their 3,600-person volunteer membership. The chapter then illustrates how one persona in particular—Wendy, the "Willing but Wary Cyclist"—was used to help crowdsource the data necessary to create a new connected network of low-stress routes in Seattle that will be usable by people of all ages and abilities.

Seattle has a historically low bicycle mode share. As well, the percentage of women, children, and elderly bicycling in the city has remained low relative to other cities that have implemented bicycle facilities designed to accommodate people of all ages and abilities. Transportation research conducted by Dill and McNeil (2012) at the Portland State University shows that bicycle facilities that cater to "vehicular cycling" (Forester 2012)—cycling in the roadway as a vehicle—feel too dangerous for most of the general population to consider using. According to Mekuria, Furth, and Nixon (2012), designing bicycle facilities for this small slice of the general population (vehicular cyclists) is an approach that will never lead to the mass cycling phenomenon seen elsewhere.

While there is a great deal of research from around the world on how to achieve "mass cycling"—cycling by a broad representation of the city demographic for basic transportation purposes—this broad body of academic transportation research does not lend itself to easy, fast, or consistent uptake by those outside of the transportation engineering, planning, and policy fields. A simpler tool was needed to make all of this cutting-edge research more intuitively digestible by the Seattle Neighborhood Greenways citizen advocates: personas.

In short, personas are fake people based on real data. Personas are tools that are commonly used by private sector commercial product design teams as a way to make large volumes of user-experience research intuitively usable by large teams of product engineers and designers who do not have research backgrounds. The end users of products are often very different from those who are designing and building them. Without a consistent user-representation in the design process, engineers and designers often have wildly differing mental models of the end users of the product. This lack of a coherent user definition usually leads to internal conflict and chaos within teams that manifests as an unsatisfactory product user experience. Given that only bicycle enthusiasts show up to bicycle planning meetings in Seattle, a new approach was needed in order to represent the needs of those who were not yet using bicycles for transportation.

Using a persona for citizen and volunteer engagement

Seattle Neighborhood Greenways is a grassroots volunteer organization created to advocate for the majority of the overall population who want safe, healthy streets that are comfortable for biking and walking for people of all ages and abilities. Impatient for consistently high quality results, Seattle Neighborhood Greenway organizers pooled their collective knowledge, experience and research to create "Wendy the Willing but Wary Cyclist" persona (Edmiston and Tuttle 2012). Wendy put a face and a name to the majority of the general population that identifies itself as being "interested but concerned" (Geller 2006, 3). According to Geller's model, interested but concerned (Willing but Wary) people are "curious about bicycling" (3) and like to ride, but are afraid to do so and therefore seldom ride and "will not venture out onto the arterials" (3).

Getting to know Wendy

Meet Wendy. Wendy is afraid to ride in Seattle traffic so she seldom gets her bike out of the garage these days. Wendy occasionally drives to the Burke-Gilman Trail to ride her bike with her kids and wishes she had safe places to ride closer to home. Wendy longs for the day when her children can ride their bikes to school safely, to soccer practice, to friends' houses, and to the library. Until then, Wendy drives her children everywhere. Since Wendy didn't grow up cycling in traffic, she never developed the skills for mixing it up with cars and trucks and probably never will voluntarily "take the lane." Wendy will only ride where she feels safe.

The Wendy demographic

Wendy's typical preferred trip is short and her favored bicycling pace is social. Most Wendys are between 35 and 54 years of age (Dill and McNeil 2012) and their level of traffic stress tolerance is low (LTS 1 and/or LTS 2).[1] Her purchase priorities related to bicycling are: comfort, affordability, and attractiveness. Wendy represents the most accessible and best source of potential new cyclists. Adding connected networks of low-stress, all ages, and all abilities bicycle facilities for the Wendys out there has resulted in large increases in the numbers of people cycling in cities that have taken this approach. Designing for Wendy will satisfy the needs of many additional potential cyclists, resulting in improved diversity metrics and public health outcomes. Cities with safe and well-planned streets have a greater range of people using bicycles.

Today, however, Wendy shops where it makes financial sense and often drives to big box stores that have plenty of free parking. However, when on her bike, Wendy shops close to home. When shopping locally by bike, Wendy visits businesses more often even though she purchases less on each visit. Wendy feels empowered and energized by her decision to shop locally by bike and she loves the lifestyle that it affords.

It would be naive to believe that many Wendys will naturally turn into Eddies (the "confident and enthused" persona) over time. Most Wendys will never become comfortable biking in situations considered to be high stress. Therefore, it is essential that Wendy perceives the low-stress bicycle facility network as connected and safe before she will use it or take her children on it. Wendy can tell you when the facility is done if she is invited to sign off on it before it's considered finished.

There are opportunities to specifically target the Wendy market segment as a massive untapped group that could help achieve ambitious mode shift targets. In order to do so, "low-stress" facility types such as protected bicycle lanes, neighborhood greenways, multi-use trails, safe arterial and bridge crossings, and other low-stress facility types proven to make Wendy feel safer, need to be prioritized over facilities that cater only to vehicular cyclists. According to research conducted by Dill and McNeil, Wendy is very concerned about being hit by a motor vehicle while bicycling. "For example, 84 percent of the Interested but Concerned group (this includes Wendy) is concerned about being hit, compared with 52% of the Enthused and Confident and 39% of the Strong and Fearless" (Dill and McNeil 2012, 7).

People like Wendy are the future of cycling in the United States. She will be the future of cycling in Seattle, too. Wendy represents the "Willing but Wary" population, which totals 60 percent of the general U.S. population. Since Wendy does not belong to bike clubs, she seldom shows up in cyclist counts, surveys, or focus groups. Because of the sampling errors inherent in current public outreach processes, Wendy's needs are seldom systematically recorded.

You might be a Wendy if:

1. You have said, "I'd love to ride a bike but the idea of riding in traffic scares me so I don't."
2. You won't let your children ride their bikes to school because you feel it isn't safe.
3. You have ridden a bike only a few times in the past 12 months, if at all. Each time, you've driven your bike to a safe place to ride or ridden it only within a few blocks of home.

4. You do not like to ride up or down steep hills. You don't have the strength to pull a trailer with two kids up steep hills.
5. You self-identify your transportation choices as "risk averse" as opposed to "excitement seeking."
6. You don't own cycling specific clothing or shoes.
7. If you own a bike, it probably has the original tires, pedals, and seat. You have not added fenders or a rack, but your bike may have come with them.
8. You chose your bike for its comfort and looks, not its components.
9. You don't know how many PSI of air are in your bike tires at this moment.
10. You have never oiled your own chain or changed a flat tire, and you don't feel comfortable adjusting your own shifters and brakes.
11. You prefer being able to reach the ground with your feet as opposed to getting "full leg extension" as your enthusiastic bike friends might say.
12. Your biking words are healthy, clean, fun, empowering, social, practical, stylish, efficient, green, frugal, kids.
13. Your biking words do NOT include Chamois Butt'r, leg extension, cadence, trigger shifter, SPD, clipless, drops, Campy, Ti, wheel suck, gear inches, *randonneur*, or index.

For Wendy, the bicycle is a short distance tool she would like to choose for local errands instead of her car. Wendy wants to use her bike for:

1. visiting friends who live in the neighborhood;
2. returning those books to the library;
3. going to the park;
4. getting to know her neighborhood businesses;
5. picnicking on the weekends with her family at different neighborhood parks;
6. getting some exercise in a way that is fun and is free;
7. short distance commuting;
8. riding with friends and their kids.

Wendy would love to teach her kids to ride bicycles like her dad did with her when she was little. She remembers learning to balance and pedaling down the street for the first time with her dad running behind, holding on to the back of her bike and then just letting go. She felt so free! Right now though, Wendy has been finding it difficult getting her children interested in any kind of regular exercise. Wendy thinks biking to the park to play might be the first step to instill healthy lifestyle habits. She wants to teach her kids family values such as self-sufficiency and autonomy.

The Wendy persona represents a broader set of people who choose to replace some of their car trips with biking and/or walking who have similar physical and psychological needs from their transportation network.

- Thirty-nine-year-old single mom Jocelyn pulls a Burley trailer with a two-year-old in it being followed by eight-year-old Sophie who has just learned how to cross intersections by herself. Sophie wants to ride her new bike to school next year. Jocelyn is wary but hopeful.
- Fifty-four-year-old Anne rides fewer than ten times per year and will never figure out how to choose the best of the 27 gears available on the bike they bought five years ago. Anne wants to go on short rides with their friends and their school-age children.
- Fifty-two-year-old Gerry has joint issues but still loves to ride. Big on smiles, but distance limited by joint pain, it is Gerry's dream to go on regular local bike rides with his two teenaged daughters before they've flown the nest for college.
- Lois is 75 years old and bikes to the grocery store now that she's successfully concluded the car ownership phase of her life. Lois can bike up to two miles from home, but can only bike up mild grades. Today, she only rides on sidewalks because she is afraid to ride in the road with fast cars in her neighborhood. Lois would like to rely on biking, walking, and bussing for all of her shopping, leisure, and social endeavors but she needs safer routes and arterial crossings. The bike is special to Lois because she feels young, strong, happy, and free on it. It also serves to keep her healthy and independent.
- Twelve-year-old paraplegic Peter uses a hand-cranked bike to get around his neighborhood. Peter also plays competitive wheelchair basketball on weekends. While endowed with superhuman enthusiasm, Peter has limited hill climbing ability and range.

Conclusions and future study

The resulting Wendy persona is a concept that the Seattle Neighborhood Greenway volunteers could identify with and advocate for. She became a unifying force in route scouting, facility evaluation, and network planning. In late 2012, 15 Seattle Neighborhood Greenway groups unified efforts to produce a citywide network of proposed low-stress routes that would be appropriate for all of the Wendys out there and their children.

This volunteer-designed route network was integrated in the draft route network for the 2014 Seattle Bicycle Master Plan (BMP) update. In fact, 95 percent of the Neighborhood Greenway routes in the adopted 2014 BMP were routes identified by Seattle Neighborhood Greenway citizen advocates. After bicycle facility implementation, the Wendy persona will be used to evaluate the quality of the user experience. "Would Wendy feel comfortable on this?" or "Would Wendy let her eight-year old daughter bike to school on her own using this route?" are typical evaluation criteria that are intuitively understandable success metrics. While Wendy has been a useful persona for advocating for those who need bicycle facilities appropriate for people of all ages and abilities, more persona development needs to be devoted to embodying the needs of the other vulnerable roadway users and modalities.

Note

1. "Level of traffic stress 1 (LTS 1) is meant to be a level that most children can tolerate; LTS 2, the level that will be tolerated by the mainstream adult population; LTS 3, the level tolerated by American cyclists who are 'enthused and confident' but still prefer having their own dedicated space for riding; and LTS 4, a level tolerated only by those characterized as 'strong and fearless'" (Mekuria, Furth, and Nixon 2012, 1).

References

Dill, Jennifer and Nathan McNeil. 2012. *Four Types of Cyclists? Testing a Typology to Better Understand Bicycling Behavior and Potential*. Portland, OR: Portland State University and the Oregon Transportation Research and Education Consortium. http://web.pdx.edu/~jdill/Types_of_Cyclists_PSUWorkingPaper.pdf.

Edmiston, Robert W. and Cathy Tuttle. 2012. *Wendy the Willing but Wary Cyclist Persona*. Seattle, WA: Seattle Neighborhood Greenways. https://docs.google.com/document/pub?id=1tv6UrIP11JLg3dHcsUex6A8fAQEdYNNoOa5so8KxP38.

Forester, John. 2012. *Effective Cycling*, Revised Edition. Cambridge, MA: The MIT Press.

Geller, Roger. 2006. *Four Types of Transportation Cyclists*. Portland, OR: Portland Office of Transportation. www.portlandoregon.gov/transportation/article/237507.

Mekuria, Maaza C., Peter G. Furth, and Hilary Nixon. 2012. *Low-Stress Bicycling and Network Connectivity*. San Jose, CA: Mineta Transportation Institute. http://transweb.sjsu.edu/PDFs/research/1005-low-stress-bicycling-network-connectivity.pdf.

4 Instagramming urban design along the Ohlone Greenway

Benedict Han

Introduction

The objective of this chapter is to evaluate the role of Instagram, a social media application, as a community engagement tool in the urban design process. This study draws upon precedents in environment-behavior studies and community engagement that have used photography as a means to gather qualitative data and gain fresh perspective directly from the users of a place. Building from these precedents, this study investigates Instagram as a community engagement tool, one that can be used to provide new types of information in conjunction with more traditional user surveys and other methods.

This investigation focused on the Ohlone Greenway, a multi-use public trail that runs through and connects the cities of Alameda and Berkeley in northern California. The trail also connects three Bay Area Regional Transit (BART) stations—North Berkeley, El Cerrito Plaza, and El Cerrito del Norte, as well as several parks and commercial districts. It utilizes the former Santa Fe railroad right-of-way, the tracks of which were laid in the late nineteenth century, connecting trade and freight between Nevada and California (Schwartz 2008).

The City of Berkeley is interested in making pedestrian and bicycle improvements to the Ohlone Greenway. The city intends to make an informed decision on specific locations to widen the pathway to a minimum of 10 feet and to improve the safety at specific intersections. From these preliminary questions, the main question driving this study emerged: what would users of the Ohlone Greenway see as top priority for improvement? Would they prioritize the same concerns that the city had? If not, what other considerations should the city make, based on community feedback?

The study methodology was based on the interface between two methods. First, an Instagram photo tour was planned with the goal to have participants take photographs of various assets and opportunities at two sites and then share the photos through Instagram for the purpose of seeing what comments might get posted. An intercept survey was also designed to serve as a control method of community engagement to see whether responses would differ from the photo tour and, if so, how. Following the execution of the tour and survey, conceptual design sections were created to compare and contrast the design recommendations coming from each method. The two different community engagement methods were purposefully separated to potentially reveal a clear distinction between the results and demonstrate how different community engagement methods can yield different outcomes related to concerns and solutions offered regarding urban design.

Results show that the Instagram photo tour can be a powerful tool in providing perspective to designers of user behavior, highlighting key assets and opportunities of the Ohlone Greenway at both sites, and overall, spatially pinpointing problems, assets, and opportunities in a space.

What is Instagram?

Instagram is a social media tool that lets users take photos, apply filters (which change the color, brightness, and various other aspects of the photos) to those photos and then share them with their "followers" along with other social media platforms such as Facebook and Twitter. Photos can be organized into groups via a hashtag (#), which allows them to be sorted by groups within Instagram for people to see and comment on. The application is free to download. Instagram creates a network for people to connect with one another; it is a platform for folks to communicate through the use of photography and allows for a user's followers to respond by giving a "like" to images they like as well as with comments and emojis.

Instagram is marketed as both a social media and a photography tool. It can be a means to bridge the gap between "concepts and models professionals use to understand and interpret reality and the concepts and perspectives of different groups in the community" (Morrow 2001, 256). The use of photography can be a means for participants to reflect on a place and its importance (Morrow 2001).

An example of this can be seen with the Photovoice method, created by Caroline Wang and Mary Ann Burris in 1997 (Kramer et al. 2012). Photovoice is defined as a "process by which people can identify, represent and enhance their community through a specific photographic technique" (Wang and Burris 1997, 369). Photovoice has three goals: "(1) To enable people to record and reflect their community's strengths and concerns, (2) to promote critical dialogue and knowledge about important community issues through large and small group discussion on photographs, and (3) reach policy makers" (Wang and Burris 1997, 370). The procedure for Photovoice goes as follows: researchers ask participants to take photos in response to a question or a specific context, for example to take a photograph of a favorite place. These photographs might be digital, though a typical technique used in Photovoice is to hand out disposable cameras that do need to be developed. Researchers then review and analyze the images, whether supplied through a medium such as Instagram or developed by the researchers themselves, and then a focus group is held with the involved community members to share the photos and debrief.

This process benefits both the participants and the researchers. For the researcher, it provides a controlled inside look at participants' perspectives of the built environment. For the participant, it provides an alternative means to get involved; the open-ended nature of photography allows for a creative means to express themselves. The Photovoice method was replicated for this study, but using Instagram as the platform for sharing and commenting on the photographs.

The Ohlone Greenway case study

Two specific study sites were identified based on existing trail conditions within the greenway. The first site was Cedar Rose Park, located at 1300 Rose Street in Berkeley. While the park has a large volume of users, the pathway itself is in poor condition; tree roots have uplifted the asphalt, making the pathway uneven and cracked. The second site is the intersection of Peralta Avenue and Hopkins Street, also located in Berkeley. At this intersection, the trail is difficult to maneuver due to the lack of views, lack of and poor quality signage, and an awkwardly shaped crosswalk. Figure 4.1 shows the site locations along the Ohlone Greenway.

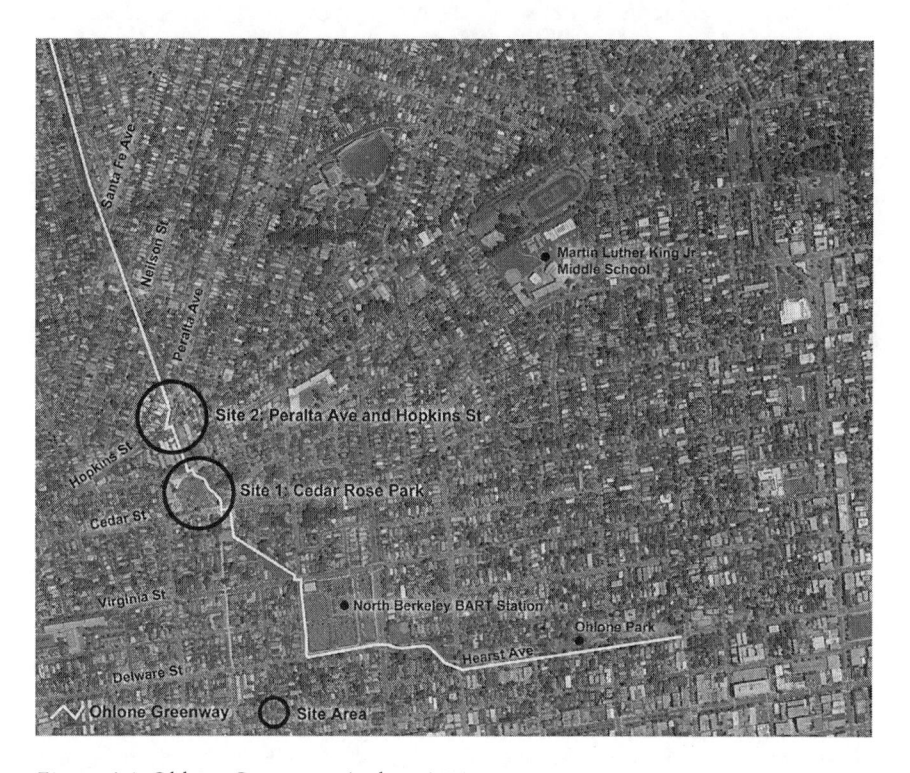

Figure 4.1 Ohlone Greenway site locations.

Existing site conditions

The study's two sites: Cedar Rose Park and the Peralta Ave. and Hopkins St. intersection are located adjacent to one another in North Berkeley. Cedar Rose Park is a large green space that is easily accessible along the Ohlone Greenway. However, the pathway here is narrow and has experienced significant wear and tear over the years. Deferred maintenance is evident by root damage to the asphalt and encroaching grass over the already narrow path. The intersection of Peralta Ave. and Hopkins St. was similarly chosen because of the challenges faced by pedestrians and cyclists alike to negotiate a difficult connection between two segments of the off-road trail. There is no clear line of sight from either end due to the configuration of the street and wayfinding clues are relatively minimal here. Both sites were chosen because they are representative of different design issues related to the Ohlone Greenway, summarized below.

Cedar Rose Park

This site was chosen due to the narrow width and poor quality of the path itself. Figure 4.2 shows Cedar Rose Park from above. The width of the pathway fluctuates between seven and eight feet wide, as shown in Figure 4.3. This creates a problem because the trail was intended to allow for traffic in both directions. The trail should be a minimum of 12' wide to accommodate two-way traffic, similar to a two-way cycle track (NACTO 2012). From site visits, it is obvious that pedestrians have to frequently step to the side to allow bicyclists to pass. For this site the hypothesis was that feedback from the survey and Instagram photo tour would show public concern that the pathway needs future improvements to allow for better circulation.

Peralta Ave. and Hopkins St. intersection

The intersection of Peralta Ave. at Hopkins St., north of the Cedar Rose Park tennis courts is very sharp, with irregularly shaped community gardens along Peralta Ave. There is no clear transition to the next segment of the off-road trail of the Ohlone Greenway. Ideally, users would come off of the greenway, which becomes a private driveway ahead, turn left onto Hopkins St., turn right on Peralta Ave. and then make another left to return to the Ohlone Greenway off-road trail. There is no clear view corridor that allows users coming from either direction to determine where the path continues. Figure 4.4 shows how the intersection blocks a clear line of sight from either end of the off-road trail. Cyclists are expected to share the road with general traffic. Figure 4.5 shows a wide neighborhood street with parking on both sides. Furthermore, there is inadequate signage to facilitate wayfinding in either direction. Finally, the private driveway heading south on Peralta Ave

Ohlone Greenway

N

Figure 4.2 Cedar Rose Park.

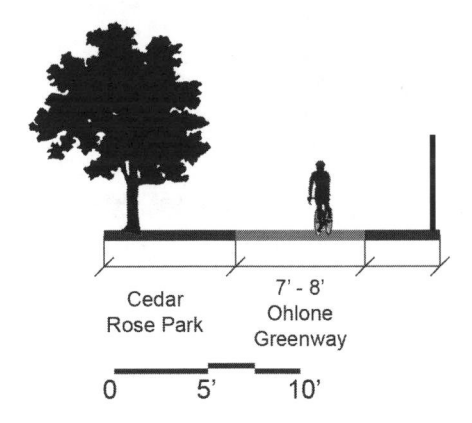

Cedar
Rose Park

7' - 8'
Ohlone
Greenway

0   5'   10'

Figure 4.3 Typical pathway section at Cedar Rose Park.

——— Ohlone Greenway

↑
N

Figure 4.4 Intersection of Peralta Ave. and Hopkins St.

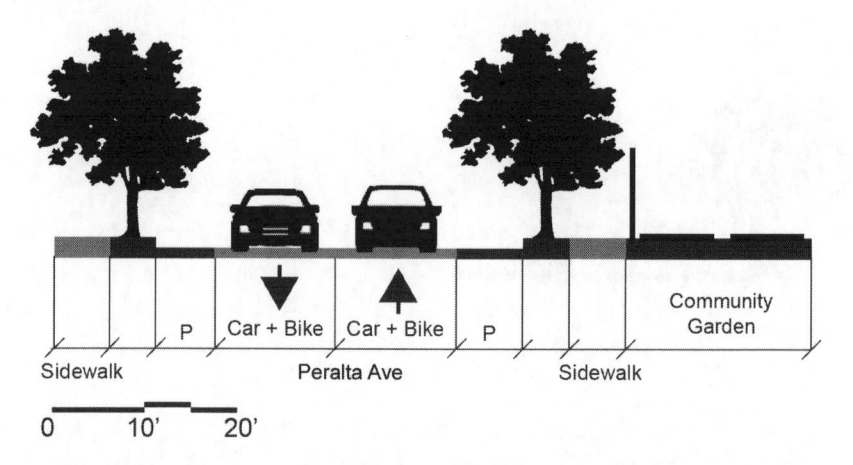

| | P | Car + Bike | Car + Bike | P | | Community Garden |
| Sidewalk | | Peralta Ave | | | Sidewalk | |

0   10'   20'

Figure 4.5 Cross section through Peralta Ave. and Hopkins St.

creates tension as this space is undefined and so shared by the homeowners and the public. The city would like to see improvements to increase way-finding and public feedback will likely show a strong dissatisfaction with the roadway crossing overall.

Methods

The Instagram photo tour conducted for this study was based off of the Photovoice method; it drew upon Instagram as a photographic tool as well as a social media platform. Participants of the tour were asked specifically to photograph assets of the trail at both study sites as well as opportunities for potential capital improvements. These questions were intentionally open-ended to give creative control to the photographer participant. An intercept survey was also designed for this study to represent the city's interest in receiving feedback on design elements that it hopes to improve. The intercept survey was conducted at both sites in one afternoon. At each site, users were asked to fill out a written survey. By comparing the two methods to one another—intercept survey and photo tour—the results revealed a difference of perspective between the city's intentions for future capital improvements and what the participant photographers deemed to be assets and opportunities related to the greenway. This juxtaposition is explored through a series of pre-conceptual designs that compare and contrast the differing results of the two different methods. More detail on each method can be found below.

Intercept survey

Intercept surveys are done to solicit feedback directly from users with regards to their experiences of a space. They are conducted in person at the location of interest for the survey. The intercept survey for this study was broken down into four parts:

- Part 1 asked questions pertaining to users' frequency of use of the Ohlone Greenway and purpose of travel.
- Part 2 asked for input on elements of the Ohlone Greenway that users liked and did not like; this section was especially critical in that it presented the same questions that were asked during the Instagram photo tour and allowed for cross comparison between methods.
- Part 3 asked users to rate various design elements of the Ohlone Greenway that the City of Berkeley identified for potential capital improvements; this section used a five-part Likert scale to reveal the relative value of each proposed capital improvement to users. The following elements of the Ohlone Greenway were evaluated in this section:

1. path width—width of the trail at this location;
2. path surface quality—how conducive is the path itself for pedestrian and bicycle circulation;
3. green/landscaped elements—what is the softscape environment surrounding the Ohlone Greenway;
4. signage—availability of signs to help promote wayfinding;
5. roadway crossings—street intersections and quality of crossing the road;
6. street furniture—objects placed along the trail (for example, benches, tables, and so forth);
7. walking/biking experience as a whole—how much users enjoy the trail at this location.

• Finally, Part 4 focused on demographics and, along with Part 1, it was used to build a profile of the users of the Ohlone Greenway.

The intercept survey was designed to serve several roles. First, it aimed to address the City of Berkeley's concerns with the Ohlone Greenway and to provide input on how to prioritize the types of improvements needed at both site locations. As introduced above, this was done through Part 3 of the survey, which asked users to rate various elements of the Ohlone Greenway at both sites. Part 3 was inspired by a study conducted by Donald Appleyard in San Francisco, looking at residents' perspectives of the environmental quality of three different types of streets (Appleyard and Lintell 1972). He asked survey participants to rate various elements of their neighborhood using a range of values to make design recommendations for improving the area. The findings from Part 3 were the primary input for creating design recommendations for both sites.

Instagram photo tour

The Instagram photo tour was done through the following steps: First, participants were taken to each site and asked to take a photo of an element of the greenway that they liked and an element that they did not like. Once a photo was taken, participants used a hashtag that I provided (#GreenwayOhlone) and wrote in the comment section why they took their photo and asked their followers to comment. Next, participants were then asked to share photos on their Facebook pages and/or Twitter accounts.

Instagrammed photos were then amassed using the #GreenwayOhlone hashtag. Finally, Instagrammed photos were coded using the themes identified from Part 3 of the intercept survey. These themes included: path width, path quality, green/landscaping elements, signage, roadway crossings, street furniture, and walking/biking experience as a whole.

Findings

From the intercept survey a total of 24 questionnaires were collected in a single afternoon. Sixty-three percent of the surveys were collected at Cedar Rose Park and 37 percent at the intersection of Peralta Ave. and Hopkins St. from respondents walking and biking in or through the sites (see Tables 4.1 and 4.4). Regarding the Instagram photo tour, 14 photos were collected from nine participants. Overall, there were a limited number of surveys and photos collected. However, the intent of this project was exploratory in nature and yielded potential refinements for future studies. The results are analyzed by site below.

Cedar Rose Park

The hypothesis for the Ohlone Greenway as it intersected Cedar Rose Park was that people would not be satisfied with the width of the path and the path quality within the site. The results from the intercept survey did indeed reveal dissatisfaction with the path, noting that it was old, which was the most common comment received, and not safe for cyclists and others to use—a condition that was significantly worsened by tree roots uplifting the path (see Table 4.2). The path surface quality was rated the lowest in the survey (see Figure 4.6). Finally, lighting was noted several times as an issue at this location.

With regard to what worked for them, respondents reported high satisfaction with the overall walking and biking experience at this site. Survey

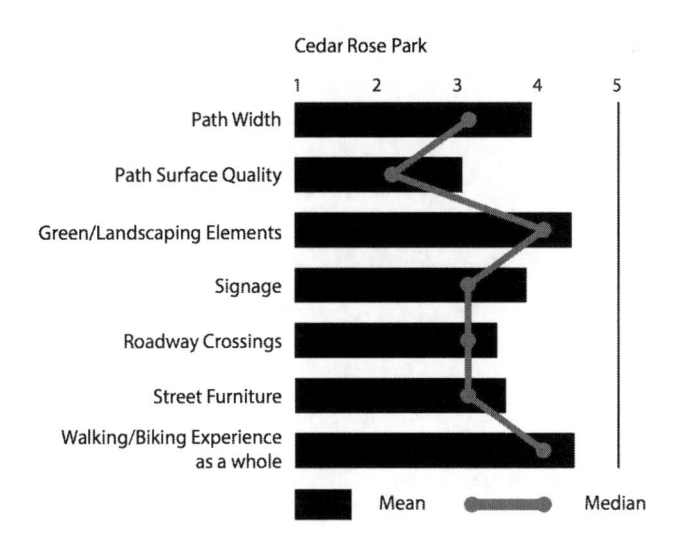

Figure 4.6 Survey participant responses from 1 to 5, with 1 being the lowest and 5 the highest.

respondents responded positively to the various aspects of the Ohlone Greenway and noted that it is a major pathway through the city and is safe to use. There was also a long list of positive adjectives that respondents used to describe the park—particularly how it is peaceful and safe. As well, people like the open space that Cedar Rose Park provides.

Table 4.1 Cedar Rose Park, survey respondent mode of transportation

Are you walking or biking?	*(%)*
Walking	86.7
Biking	06.7
Both	06.7

Where did you start your current trip?	*(%)*
Home	80.0
Work	00.0
BART	06.7
Other	13.3

What is your current destination?	*(%)*
Work	00.0
Home	66.7
Ohlone Park	0.00
Cedar Rose Park	06.7
North Berkeley BART Station	06.7
Retail or restaurant	00.0
Other	20.0

If you're going to Cedar Rose Park, what is your destination within the park?	*(%)*
Tennis court	00.0
Basketball	06.7
Playground	26.7
Community Center	00.0
Lawn	06.7
Other	00.0

How frequently do you use the Ohlone Greenway?	*(%)*
Greater than 5 times a week	20.0
3 to 5 times a week	13.3
1 to 2 times a week	20.0
1 to 2 times a month	46.7

What is your most frequent way of using the pathway?	*(%)*
Walking	66.7
Biking	13.3
Both	20.0
Other	00.0

Table 4.2 Cedar Rose Park, survey respondent likes and dislikes

What do you like most about the Ohlone Greenway?	
Path width	Main path through Berkeley
Path surface quality	Fresh air, Open space
Green elements	
Signage	
Roadway crossings	
Street furniture	
Walking/Bike experience	Pretty, Fun, Quiet, Kid-friendly, Clean, Dog-friendly, No cars, Peaceful, Safe
Other	

What do you not like about the Ohlone Greenway?	
Path width	Narrow path
Path surface quality	Old path, Overall quality, Bad for biking, Squishiness
Green elements	
Signage	
Roadway crossings	
Street Furniture	No sitting space
Walking/Bike experience	Maintenance, Cars, Dark, Lack of light
Other	No basketball court

Table 4.3 Cedar Rose Park, survey respondent demographics

Age group?	(%)
18–24	40.0
25–34	13.3
35–49	40.0
50–64	13.0
65+	00.0
Did not answer	06.7

Race	(%)
Caucasian	46.7
African American	13.3
Hispanic/Latino	13.3
Asian	13.3
Pacific Islander	00.0
Native American	00.0
Other	06.7
Did not answer	06.7

Sex	(%)
Male	33.3
Female	60.0
Non-binary	00.0
Did not answer	06.7

Figure 4.7 Examples of Instagram photos posted in response to the question: "What do you see as an asset to this site?"

Figure 4.8 Examples of Instagram photos highlighting potential improvements posted in response to the question: "What do you see as an opportunity for this site?"

Photos from the Instagram photo tour were organized into different themes based on the seven design elements used in Part 3 of the intercept survey. A total of seven photos were collected for this site. All of the photos at Cedar Rose Park (see Figures 4.7 and 4.8) emphasized either the street furniture, green elements or signage. Comments to each photo are shown below the photo as well as the number of "likes" from their followers.

Other photos Instagrammed as part of the photo tour showed green elements and street furniture to be key assets of the Ohlone Greenway at Cedar Rose Park. These results aligned with the survey findings, which also highlighted these elements.

Peralta Ave. and Hopkins St.

The hypothesis at this site was that there would be the most dissatisfaction with the intersection of Peralta Ave. and Hopkins St., based on a lack of effective wayfinding in this area. The results from the survey for this site indicated that there is a strong dissatisfaction with both the roadway crossings and signage. The most frequent response was the lack of lighting on the path—a similar issue at Cedar Rose Park. There were also comments regarding the lack of sufficient green elements in the area and a desire for a water feature. The results, as shown in Figure 4.9, show that the roadway crossings and signage are rated the lowest of all the criteria. There were also several complaints about the signage—there was one comment about the difficulty of way finding and graffiti on lampposts. However, when respondents were asked about their most and least favorite aspects of this site (see Table 4.5), there was significantly less focus on such elements. Instead, the comments were focused on the lack of specific street furniture, such as benches and lighting, something that was previously not considered by the City of Berkeley.

Overall, the results of Part 3 of the survey at Peralta Ave. and Hopkins St. align with the City's wishes to improve wayfinding in this area along with making improvements to this intersection to facilitate circulation in the area. Looking at what survey respondents liked the most, the results show that, similarly to Cedar Rose Park, they enjoyed how the path provides a shortcut through the City of Berkeley and how peaceful the space is. Furthermore, the presence of greenery was very appreciated. The artwork on Peralta Ave., a feature not found at Cedar Rose Park, was also mentioned once. Most of the positive comments on the Ohlone Greenway were on peoples' walking and biking experiences as a whole. There were no positive comments on the road crossings or the signage.

A total of seven photos were taken at this site during the Instagram photo tour. As with Cedar Rose Park, all the photos emphasized street furniture and green/landscaping elements. However, a key finding from these photos was that participants highlighted the difficulties of crossing the intersection of Peralta Ave. and Hopkins St.; the private driveway that intersects with the Ohlone Greenway was identified as a key issue.

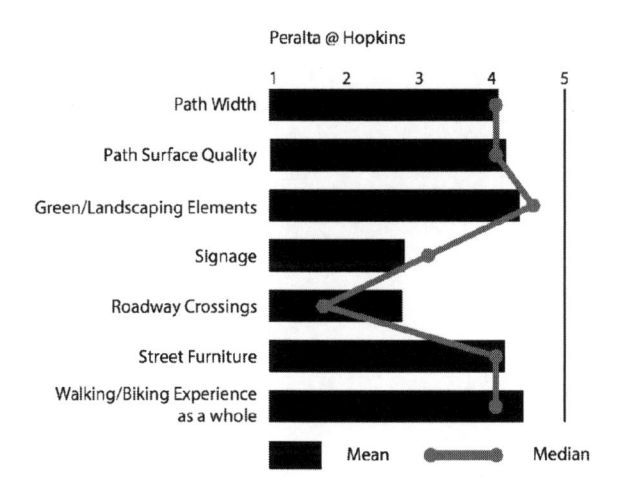

Figure 4.9 Survey participant responses from 1 to 5, with 1 being the lowest and 5 the highest.

All of the Instagram photos (see Figure 4.10) identified as an asset to the Ohlone Greenway at Peralta Ave. and Hopkins St. show the street furniture and green elements as key assets. This comes as no surprise as there are wayfinding kiosks and public artwork here as a part of the site area.

Table 4.4 Peralta Ave. and Hopkins St., respondent mode of transportation

Are you walking or biking?	*(%)*
Walking	77.8
Biking	22.2
Both	00.0

Where did you start your current trip?	*(%)*
Home	88.9
Work	00.0
BART	00.0
Other	11.1

What is your current destination?	*(%)*
Work	00.0
Home	55.5
Ohlone Park	00.0
Cedar Rose Park (CRP)	00.0
North Berkeley BART Station	00.0
Retail or restaurant	11.1
Other	11.1

If you're going to CRP, what is your destination within the park?	*(%)*
Tennis court	00.0
Basketball	00.0
Playground	11.1
Community Center	00.0
Lawn	00.0
Other	11.1

How frequently do you use the Ohlone Greenway?	*(%)*
Greater than 5 times a week	22.2
3 to 5 times a week	33.3
1 to 2 times a week	11.1
1 to 2 times a month	22.2

What is your most frequent way of using the pathway?	*(%)*
Walking	44.4
Biking	33.3
Both	11.1
Other	11.1

Table 4.5 Peralta Ave. and Hopkins St., survey respondent likes and dislikes

What do you like most about the Ohlone Greenway?	
Path width	Off the grid
Path surface quality	
Green elements	
Signage	
Roadway crossings	
Street furniture	
Walking/Bike experience	Quiet, Nice way to be in contact with people, Everything
Other	Artwork

What do you not like about the Ohlone Greenway?	
Path width	
Path surface quality	
Green elements	No water feature, Grass, Not enough greenery
Signage	Hard to find path, Graffiti on lampposts
Roadway crossings	
Street furniture	No lights, Not enough benches
Walking/Bike experience	Cars, Construction, Garbage, Noise when closed
Other	

Table 4.6 Peralta Ave. and Hopkins St., survey respondent demographics

Age Group?	(%)
18–24	00.0
25–34	11.1
35–49	33.3
50–64	33.3
65+	11.1
Did not answer	11.1

Race	(%)
Caucasian	55.5
African American	11.1
Hispanic/Latino	11.1
Asian	00.0
Pacific Islander	00.0
Native American	00.0
Other	00.0
Did not answer	22.2

Sex	(%)
Male	55.5
Female	33.3
Non-binary	00.0
Did not answer	11.1

Figure 4.10 Examples of Instagram photos posted in response to the question: "What do you see as an asset to this site?"

Figure 4.11 Examples of Instagram photos highlighting potential improvements posted in response to the question: "What do you see as an opportunity for this site?"

Opportunities for improvement along this part of the Ohlone Greenway centered around the lack of lighting and safety concerns brought up by photo tour participants, as shown in Figure 4.11 above.

Focus group: post project review

After the Instagram photo tour, a focus group was conducted with the participants to get their impressions of the process. Overall, the Instagram photo tour participants enjoyed participating in the photo tour because of the challenge of capturing their ideas for a social media post. A key question asked in the focus group was whether or not the participants had the same concerns as the City of Berkeley when looking at the Ohlone Greenway at both sites (i.e. path widening and safety). The participants responded that they did not notice these elements. Rather, they were more focused on the use of space, as shown in the photos. Furthermore, with regards to the types of photos, only two photos reflected concern with the pathway itself. Participants were not briefed on the city's ideas for capital improvements for the Ohlone Greenway prior to the Instagram tour. This was intentionally done to prevent any influence in their exploration of space. For future studies, it would be interesting to explore the impact of a prior orientation for participants on photos taken for research. Finally, when asked what types of groups would be most interested in participating in future Instagram photo tours, youth were a key demographic where participants found consensus.

A key challenge noted was that the participants related the difficulty of visually representing concepts specific to the Ohlone Greenway. For example, several participants wanted to highlight the lack of a view corridor as one comes onto Hopkins St. due to the community garden. Another key challenge identified was that in a photo tour, participants could not visually represent the larger picture of the Ohlone Greenway or express more regional assets—only specific elements at each site. For example, a favorite characteristic of the Ohlone Greenway from the intercept survey was how it provides direct access and mobility throughout the City of Berkeley—something like this would have been difficult to capture in an Instagram photo.

Key findings from the photo tour

The majority of photos emphasized user behavior with the various design elements at each site. The photos from the Instagram photo tour added a new perspective from users without any critique of how space is used. As mentioned from the focus group with the photo tour participants, the photos that they took were focused on the use of space more than from an engineering or planning perspective, which was focused on the pathway and its condition. However, a key challenge was how photography can be limited in expressing ideas of why a space is an asset. For example, it is difficult to explain how the Ohlone Greenway as a direct pathway through the City of Berkeley is a benefit in a photo. Future studies should explore using online platforms that can accommodate supplemental information, such as notes and maps.

Another challenge that arose from the photo tour was that some of the participants had privacy settings switched on in their Instagram accounts. They posted their #GreenwayOhlone photos, but because of the privacy settings, it limited who could comment. Also, the privacy setting could not be changed after the fact. The appropriate setting for photos posted during the tour would have been "public." Because of this not all of the photos could be used in the analysis, something that should be corrected in future studies.

The photos focused on aesthetically pleasing elements, such as public art or a flowering tree. Photo tour participants responded well to elements in each site that they thought had strong aesthetic value. This was particularly true at the Peralta Ave. and Hopkins St. intersection site, where there are public art installations in the form of benches and other street furniture. The idea of the role public art can play in a pedestrian or biker's experience along the trail was not once brought up in the survey, which was somewhat surprising. Public art can play a role in increased pedestrian and cyclist activity in the city, and is even an indicator of quality pedestrian environments according to the San Francisco Department of Public Health (San Francisco Department of Health 2013).

Instagram photos provided analysis of how elements are seen as both assets and opportunities for additional improvement. Both methods of engagement—survey and photo tour—noted street furniture and green elements with a positive rating, in the case of the intercept survey, or as an asset, as noted by the photo tour. The Instagram photo tour process allowed respondents to dive deeper into evaluation and provide examples of what works and what has room for improvement. For example, for street furniture the photos showed which specific amenities in the park were more favorably viewed. However, some of the same things that were rated highly were also noted as an opportunity for improvement. An example of this can be seen with the evaluation of green elements at the Peralta Ave. and Hopkins St. site. See Figure 4.12 for an example of each. The image on the left shows a green element that a participant liked versus an image on the right that shows an opportunity for development at the Peralta Ave. and Hopkins St. site. The intercept survey rated green elements a 4 out of 5. Both methods of engagement noted street furniture and green elements with a positive rating, in the case of the intercept survey, or as an asset, as noted by the photo tour. The survey was successful in that it provided a quantitative value to each category surveyed, which allowed for comparison between both sites and various elements at each but lacked specifics on why different elements were rated.

In comparing the results of the Instagram photo to Part 2 of the intercept survey (What did you like/dislike the most about this site), the answers had significant overlap. Part 2 of the intercept survey was designed to ask the same questions as the Instagram photo tour by asking participants to write what they liked and did not like at each site. The results from this section were very similar to the photos from the Instagram photo tour, though provided through different means. The similarity of results shows promise in the role of Instagram as a platform for community engagement.

Figure 4.12 Two Instagram photos showing how green elements (front and side yards) are an asset and also an opportunity for additional improvement.

Design recommendations

The final step of this study was to see how the results from each method by themselves influence design proposals for the Ohlone Greenway at Cedar Rose Park and the intersection of Peralta Ave. and Hopkins St. Table 4.7 shows the design considerations for each site based on the recommendations from both methods of community engagement used in the study. The pre-conceptual designs made from the intercept survey and the Instagram photo tour should be seen as a visualization of the results only, not as a recommendation to the city. Results from the intercept survey and the Instagram photo tour were kept separate to visually demonstrate that the community engagement strategies employed by a city or a designer can have an impact on the initial design concepts made for changing the public realm.

Cedar Rose Park conceptual design

Figures 4.13 and 4.14 showcase a pre-conceptual design based on the results of the intercept survey and the Instagram photo tour respectively. The key difference between the two sets of engagement methods was the use of space. The intercept survey highlighted a need to improve the Ohlone Greenway to be more conducive for active transportation modes. Figure 4.13 shows a widened pathway. In contrast, the Instagram photo tour highlighted a need to improve use of space by improving lighting and providing more street furniture. Figure 4.14 shows the inclusion of pedestrian lighting as a form of improved lighting and additional site furniture to further activate the space.

Peralta Ave. and Hopkins St. intersection conceptual design

Figures 4.15 and 4.16 show the conceptual designs for the Intersection of Peralta Ave and Hopkins St. From the intercept survey, the roadway crossing and signage were brought up as major issues at this site—there is no clear, dedicated pathway for cyclists. One example of this can be seen through the addition of a cycle track by removing the parking lanes. Separating the use of the space and adding a bicycle facility would increase the visibility

Table 4.7 Design recommendations according to method

	Intercept survey	*Instagram photo tour*
Cedar Rose Park	• Widen pathway to 12'	• More light and street furniture
Peralta Ave. and Hopkins St.	• Protected bike lane to increase cyclist visibility	• Separate private driveway
	• More signage	• Wayfinding

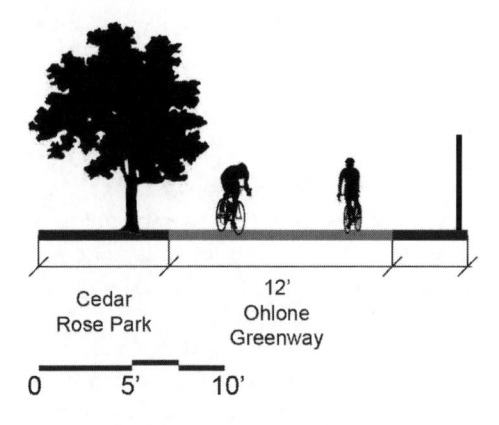

Figure 4.13 Design considerations from the intercept survey at Cedar Rose Park.

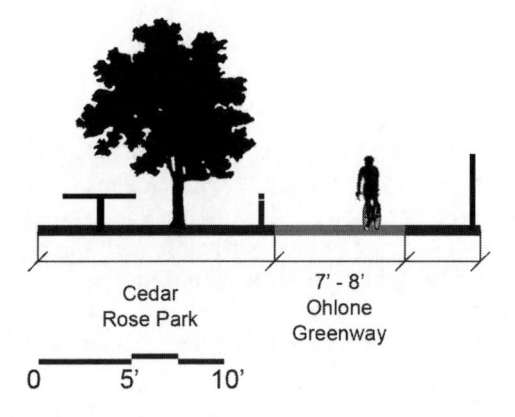

Figure 4.14 Design considerations from the Instagram photo tour at Cedar Rose Park.

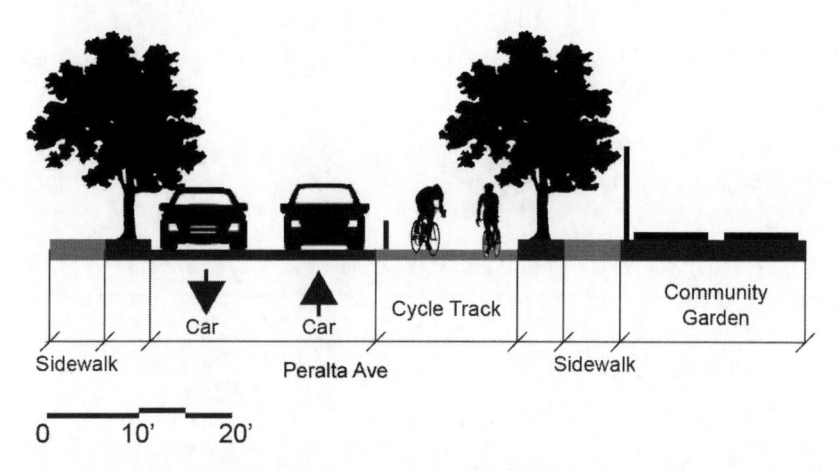

Figure 4.15 Design considerations from the intercept survey at Peralta Ave. and Hopkins St.

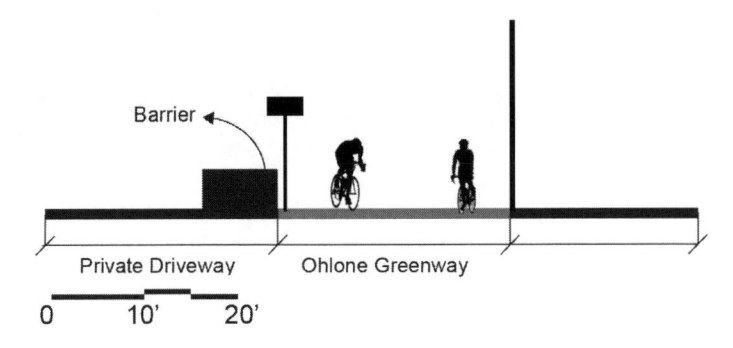

Barrier

Private Driveway   Ohlone Greenway

0   10'   20'

Figure 4.16 Design considerations from the Instagram photo tour at Peralta Ave.
and Hopkins St.

of the bike path from either direction of the Ohlone Greenway at this location. With regards to the Instagram photo tour, a key need brought up was to address the driveway on the south side of Hopkins St. The driveway is for a private residential development west of the Ohlone Greenway but often becomes a place of conflict for cyclists and cars. Separating the use of space, even by means of a planter, is an easy capital improvement to improve the safety of cyclists at this location (see Figure 4.16).

Conclusions

In terms of the effectiveness of Instagram, all of the photos highlight the use of space, and provide a means for designers to spatially pinpoint opportunities within a site. This can also be seen with Part 3 of the intercept survey, which duplicated the questions of the photo tour in written format. The similarity of answers between the Instagram photos and intercept survey begs the question of whether or not Instagram can be an effective method of community engagement. On one side, the fact that very similar answers can be found from two different methods could validate the idea that Instagram is one means by which data can be gathered. On the other side, asking a question using the photo tour format rather than a survey could be a less time-intensive, more effective method for community engagement. However, when directly asked about the issues with each site, as seen in Part 3 of the Intercept survey, there was some consensus between the city and survey respondents. In terms of next steps, it would be interesting to see the types of photos that emerged when participants were introduced to a city's concern for a space prior to doing the Photo tour.

Despite the findings, there were a few challenges and shortcomings. First, there were limited participants and photos in both methods of engagement. Second, there are some limitations to the act of photography also. Participants in the focus group noted that they had a hard time visually representing certain concepts. For example, several noted that it would be an opportunity to create a view corridor at the intersection of Peralta Ave. and Hopkins St. as you head north to provide better wayfinding along the Ohlone Greenway. Furthermore, there was no way to easily mention regional issues with the Photo tour. This can be seen in Part 2 of the intercept survey, which shows that a clear asset of the space was how it provides direct access and a short cut to different neighborhoods in the City of Berkeley. However, this was not shown at all in the Instagram photo tour. Finally, there is less control over the answers. While asking specific questions about topics of interest can mitigate this, more direction will be needed for more specific answers. Also, any researcher or planner who uses this method needs to ensure that they have full access to the photos, including Instagram accounts and other social media that may be used in this process.

Based on this study, Instagram does have the potential to be used as an initial needs assessment tool in identifying assets and opportunities for improvement in a space. The intercept survey and Instagram photo tour highlight needs beyond the path width and quality at Cedar Rose Park and the roadway crossing and signage at the intersection at Peralta Ave. and Hopkins St. The biggest take away from this process was how participants of the Instagram tour genuinely had fun. For designers and city officials, there is significant potential for this process to become a community engagement tool that helps to engage people outside of public meetings and focus groups. For future consideration, it would be interesting to see the role this process can play in building relationships between community stakeholders with each other, city officials, and planners. As the world becomes more dependent on the Internet and as social media becomes more prevalent in people's lives, Instagram can prove be a new venue for engagement. It should not replace real-time interactions, but it can be a means to bridge these two worlds.

Acknowledgments

I would like to express my appreciation and gratitude to all those who provided me the opportunity to write this chapter. First, I would like to thank Eric Anderson, Associate Planner with the City of Berkeley, for the chance to work with the City of Berkeley. I would also like to thank UC Berkeley professors Elizabeth Macdonald and Peter Bosselmann for their support in developing the study. Finally, I want to thank the participants of my Instagram photo tour and the intercept survey respondents for providing valuable data and insights.

References

Appleyard, Donald and Mark Lintell. 1972. "The Environmental Quality of City Streets: The Residents' Viewpoint." *Journal of the American Institute of Planners* 38(2): 84–101.

Kramer, Leila, Pamela Schwartz, Allen Cheadle, and Suzanne Rauzon. 2012. "Using Photovoice as a Participatory Evaluation Tool in Kaiser Permanente's Community Health Initiative." *Health Promotion Practice* 14(5): 686–694.

Morrow, Virginia. 2001. "Using Qualitative Methods to Elicit Young People's Perspectives on their Environments: Some Ideas for Community Health Initiatives." *Health Education Research* 16(3): 255–268.

NACTO (National Association of City Transportation Officials). 2012. *Urban Bikeway Design Guide*. New York, NY: NACTO. http://nacto.org/publication/urban-bikeway- design-guide/cycle-tracks/.

San Francisco Department of Public Health. Program on Health, Equity and Sustainability—Pedestrian Environmental Quality Index. www.sfhealthequity.org/elements/24-elements/tools/106-pedestrian-environmental-quality-index.

Schwartz, Susan. 2008. *The Santa Fe Right of Way in Berkeley*. Friends of Five Creeks. www.fivecreeks.org/history/walkSFROWMay08.pdf.

Wang, Caroline and Mary Ann Burris. 1997. "Photovoice: Concept, Methodology, and Use for Participatory Needs Assessment." *Health Education & Behavior* 24(3): 369–387.

5 A look at bicycle commuting by low-income New Yorkers using the CEO Poverty Measure

*Todd Seidel, Mark Levitan,
Christine D'Onofrio, John Krampner,
and Daniel Scheer*

Introduction

Commuting by bicycle in New York City increased by nearly 90 percent from 2005 to 2011 according to the Census Bureau's American Community Survey (ACS). The common perception is that bicycle commuters are white and middle to upper class. Data suggest otherwise. According to the ACS, in 2011 over 45 percent of those commuting by bicycle were non-white and close to 30 percent were below 150 percent of the poverty threshold, capturing those in poverty and near-poverty, based on the New York City Center for Economic Opportunity (CEO) Poverty Measure.[1]

This chapter uses the CEO Poverty Measure to explore demographic characteristics and incomes of bicycle commuters below 150 percent of the poverty threshold in New York City between 2005 and 2011. It also compares the locations of neighborhoods where low-income bicycle commuters live and work to the existing bicycle infrastructure, which the New York City Department of Transportation (NYCDOT) has expanded by over 100 percent in the past five years. Use of this expanding bicycle infrastructure by the working poor and those vulnerable to poverty promises to have not only positive health and environmental benefits but also positive impacts on household budgets by lowering commuting costs. The NYCDOT has an ambitious goal of tripling the number of bicycle commuters by 2020; understanding more about low-income New York bicycle commuters may help them and future low-income bicycle commuters benefit economically from NYCDOT investments.

Background

New York City is in the midst of a bicycling infrastructure boom. In the past decade, the NYCDOT has added over 300 miles to the bike lane network and installed over 8,000 new bike racks (NYCDOT 2012, 2013).

In 2009, the City Council passed a law to improve conditions for bike commuters (NYCDOT 2009) and New York City is now rolling out its ambitious bike share program—Citi Bike—with an anticipated build-out of 10,000 bikes and 600 stations (Citi Bike 2013).[2]

Along with the infrastructure improvements, the use of bikes in New York has also experienced an unprecedented rise. Both local and federal data indicate significant increases in the use of bikes for leisure trips and commuting in the city. The NYCDOT shows a 250 percent increase in its annual bike count from 2000 to 2012 (NYCDOT 2013a) and the ACS indicates that from 2005 through 2011, commuting by bike increased almost 90 percent in New York City.[3]

Even with these impressive increases, the improvements in bicycle infrastructure have not been free from criticism. Bike lanes have been called a pathway to gentrification and the whim of a totalitarian and autocratic leader (Applebaum et al. 2011; Wall Street Journal 2013). As Mayor Bloomberg's final term came to a close there was anxiety about the future of bike lanes. Some mayoral candidates indicated that they would strongly consider removing existing bike lanes (Flegenheimer 2013).[4] Such criticism has not been isolated to New York. In Washington, D.C. and Portland, OR, concerns have been raised that the expanded bicycle infrastructure overwhelmingly benefits white, affluent residents (Davis 2011; Tavernise 2011).

However, data suggest otherwise. The ACS indicates that over 45 percent of New York City bike commuters are of a race/ethnicity other than non-Hispanic (NH) white. Close to 30 percent of all bike commuters are under 150 percent of the poverty threshold, capturing those in poverty or near-poverty, as set by New York City's CEO Poverty Measure, an alternative poverty measure pioneered by New York City and developed by this chapter's authors under the leadership of Dr. Mark Levitan.

Adopted in 2008 by New York City as the first alternative poverty measure in the nation, the CEO Poverty Measure includes the effects of taxes and the value of in-kind assistance (such as nutritional assistance and housing subsidies) on family resources.[5] It also deducts from family resources non-discretionary spending for items such as commuting costs, childcare and out-of-pocket medical expenses to widen the definition of family resources and create a more realistic measure of what is available for families to purchase necessities. The result is a broader, more realistic measure of poverty with the ability to measure the effectiveness of specific policy programs in lowering the poverty rate.

Most relevant to our discussion here is the low cost of commuting by bicycle and the substantial positive impact on the pocketbooks of those that choose to do it. A recent report by the Mineta Transportation Institute indicates that low-income households are concerned with their transportation costs and make transportation decisions based on the cost of each mode of travel (Mineta 2011). As one of the most affordable commuting methods available, the advantages of commuting by bicycle are quantifiable: in 2011,

CEO estimates indicate that New York City bike commuters spent annually $972 less than the average subway or bus commuter and $2,417 less than the average motorist. This chapter examines these monetary benefits, along with the demographic and economic characteristics of bike commuters in New York City.

We start by explaining the need for the CEO Poverty Measure. This section provides a brief description of the shortcomings of the Official Poverty Measure that led to the development of an alternative method. We also describe the methodology used to calculate the CEO's costs related to commuting.

The second section explores the demographics and economic character-istics of the city's bike commuters. Recent literature has coined the term "invisible cyclist" for low-income and, predominantly Hispanic, immigrant populations that have turned to using bikes as a necessary cost-effective method of transportation (Applebaum et al. 2011; Fuller and Beltran 2010; Huff 2011; Leung and Mannos 2011; Smart 2010). They are described as such because they are a cycling population with limited advocacy–are "in-visible" to planning decisions. We examine the data to see whether the "invisible cyclist" population is represented.

In the last section of this chapter we present a number of hypothetical scenarios to examine the benefits of bicycle commuting. Currently, less than 1 percent of all New York City commuters use bikes for their journey to work. Due to the cost-effectiveness of bike commuting, if this fraction increased, we would see quantifiable impacts on incomes and consequen-tially on the CEO poverty rate. We will examine the impact on the CEO poverty rate assuming New York attained the bike-commuting rate of two of the U.S. cities with the highest share of bike commuters, namely, Portland, OR, and Seattle, WA, and a somewhat ambitious bike-commuting rate of 15 percent of total trips.

The Official and CEO Poverty Measures

It has been over a half century since the introduction of the current official measure of poverty used in the United States, the Official Poverty Measure. In the early 1960s, this measure represented an important advance, serving as a focal point for the public's growing concern about poverty in America. But over the decades, discussions about poverty have increasingly included criticisms of how poorly it was being measured and, therefore, understood.

The Census Bureau's Official Poverty Measure includes only pre-tax cash in its definition of resources that are compared against the poverty threshold. In recent years, an increasing share of what government programs do to support low-income families takes the form of tax credits (such as the Earned Income Tax Credit) and in-kind benefits (such as Food Stamps).[6] Therefore the official measure does not account for how these programs provide resources to individuals and families, nor how these programs affect poverty. Nor does it account for the impact of non-discretionary spending on commuting costs, childcare and medical out-of-pocket expenses on the family budget.

Dissatisfaction with the Official Poverty Measure prompted Congress to request a study by the National Academy of Sciences (NAS). The NAS's recommendations, issued in 1995, sparked further research and garnered widespread support among poverty experts (Citro and Michael 1995). However, neither the federal nor any state or local government had adopted the NAS approach until Mayor Bloomberg commissioned the Center for Economic Opportunity to develop a poverty measure that would be more responsive to policy changes. In August 2008, under the leadership of Mark Levitan, CEO released its initial report on poverty (Levitan et al. 2008) highlighting the development of the CEO Poverty Measure.

The NAS recommendations, which CEO adopted, took a considerably different approach to calculating both the threshold and resource sides of the poverty measure (Levitan et al. 2013).[7] The threshold reflects the need for clothing, shelter, and utilities as well as food. It is established by selecting a sub-group of families as reference families,[8] calculating their spending on these items, and then choosing a point in the resulting expenditure distribution.[9] A small multiplier is applied to account for miscellaneous expenses such as personal care, household supplies, and non-work-related transportation. The threshold is updated each year by the change in the level of this spending.[10] This connects the threshold to changes in living standards. In further contrast to the official measure, the CEO measure is adjusted to reflect the relatively high cost of living for New York City.[11]

On the resource side, the CEO measure is designed to account for the flow of income and in-kind benefits that a family can use to meet their basic needs. This creates a much more inclusive measure of income than simply looking at pre-tax cash. The tax system and the cash-equivalent value of in-kind benefits for food and housing create important additions to family resources. But families also have non-discretionary expenses that reduce the income available to meet their other needs. These include the cost of child-care, commuting to work, and medical care that must be paid for out-of-pocket. This non-discretionary spending is accounted for as deductions from income. We call this aggregate level of income available to meet family needs, CEO income. A family's CEO income level is then measured against the family's threshold to determine the family's poverty status.

Most important to this chapter is the deduction of the cost of commuting from CEO income. NAS recommendations were to deduct the cost of commuting because these are expenses necessary to earn income. Therefore the money spent commuting was unavailable to families to meet their basic needs of food, clothing, and shelter (plus miscellaneous spending) represented in the threshold amount. CEO has adopted that recommendation in its methodology.

To measure the resources available to a family to meet the needs represented by the threshold, our poverty measure employs the Public Use

Microdata Sample (PUMS) from the Census Bureau's American Community Survey (ACS) as its principal data set. The ACS is designed to provide measures of socioeconomic conditions on an annual basis in states and larger localities. It offers a robust sample for New York City (roughly 25,000 households) and contains essential information about household composition, family relationships, and cash income from a variety of sources.

Unfortunately, the ACS provides only some of the information needed to estimate these additional resources. But, as noted earlier, the NAS-recommended poverty measure greatly expands the scope of resources that are measured in order to determine whether a family is poor. CEO has developed a variety of models that estimate the effects of commuting costs, taxation, nutritional and housing assistance, work-related expenses, and medical out-of-pocket expenditures on total family resources and poverty status. More details on how these non-cash income items are estimated are available in our latest report.[12] However, due to its relevance to this chapter, the next section explains the methodology we use to calculate the cost of commuting.

Estimating commuting costs

To estimate commuting costs, we use the available ACS variables and relevant administrative data. The ACS provides data on the means of transportation to work, the travel time, usual weekly hours, vehicle occupancy, work location, and weeks worked in the past 12 months. Using these ACS variables, we establish the means by which the journey to work was made. We then use administrative data to establish a cost per trip. Once we have established a cost per trip for each means of transportation (other than railroad, which is already calculated as a weekly expense in the ACS), we use the formula below to calculate the weekly commuting cost:

$$\text{Weekly Commuting Cost} = (\text{Cost/Trip} \times \text{Min}((\text{WKHP}/8 \times 2),14))$$

We assume an eight-hour workday and use the ACS variable "WKHP—Usual hours worked per week in the past 12 months" to calculate the number of days worked per week.[13] To account for a trip to and from work, we then multiply the number of workdays by two and cap the number of possible weekly trips at 14. The cost per trip is then multiplied by the number of commuting trips per week to establish a weekly commuting cost. This is then multiplied by the "WKW—Weeks worked in the last 12 months"[14] to establish the annual commuting cost.

We have grouped bike commuting with walking and working at home, which have a per trip cost of $0. While researching this chapter, we recognized that there are initial costs and ongoing maintenance costs associated with bike commuting. Both the League of American Bicyclists[15] and the Bicycle Commuter Act, which became law in January of 2009 estimate an annual cost of between $240 and $300 a year.[16] As a comparison, the IRS Standard Mileage Rate (SMR), which we use to estimate commuting costs for automobiles is based on the fixed (vehicle purchase or lease, registration and insurance) and variable costs (maintenance, repair, fuel and oil) of operating an automobile (Victoria 2011). Therefore, in future revisions of the CEO Poverty Measure, we will look at modifying our cost for commuting via bicycle to include bike maintenance, bike parts, and wear-and-tear cost of commuting, most likely using a value similar to those established by the League of American Bicyclists and the Bicycle Commuter Act. However, for this analysis we stuck with the $0 per trip for bike commuting.

Demographics of New York City bike commuters

As mentioned earlier, the numbers of cyclists and the amount of bicycle infrastructure in New York City have expanded rapidly in recent years. The New York City DOT reports that cycling has increased by 350 percent since 1990, according to their Bicycle Screenline Counts.[17] Unfortunately, due to limitations in the ACS, the data reported in this chapter do not reflect these actual increases in City bike trips but only the increases in those responding to the survey that their method of commuting was a bicycle.[18] Regardless of these drawbacks, the ACS still records a 90 percent increase in bicycle commuters from 2005 to 2011, as shown in Figure 5.1.

Even with the increase, the fraction of New York commuters that use a bike is still under 1 percent, a fairly small, but growing, percentage of all commuters.[19] Therefore to facilitate analysis of more granular data, such as demographic characteristics and income data, we have compiled a five-year ACS dataset. We use this five-year dataset, called "2007–2011 American Community Survey Public Use Microdata Sample, Augmented by CEO," for our demographic and income analysis. It is important to note that by

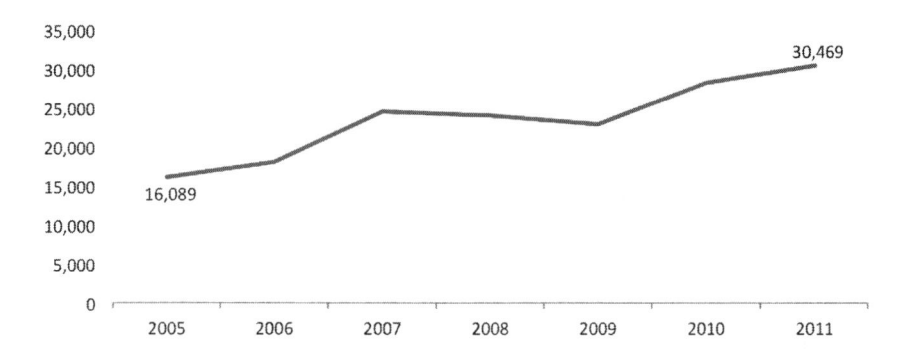

Figure 5.1 New York City bike commuters.
Source: 2005–2011 American Community Survey Public Use Microdata Sample.

using the five-year dataset we are unable to look at year-to-year changes among these more detailed groups.

In Tables 5.1, 5.2, and 5.3, we report the demographic and income groups of all bike commuters. We also report the data for all New York City commuters. We feel that it is important to have this comparative analysis to put the bike commuting data in relative perspective.

The ACS data indicates that Hispanics represent almost 30 percent of all New York City bike commuters. This is about 4 percentage points higher than their share of all New York City commuters. Hispanics and non-Hispanic white (NH White), who make up almost 55 percent of all bike commuters, are the only two race/ethnicity groups that have a higher relative share of bike commuters than all commuters. Together they represent almost 85 percent of bike commuters while representing 63.5 percent of all commuters. The other three race/ethnicity groups—non-Hispanic black (NH Black), non-Hispanic Asian (NH Asian), and Other race/ethnic group—make up only 15.5 percent of all bike commuters while representing 36.5 percent of all commuters.

For this analysis, we aggregated educational attainment into two groups, "high school or less" versus "some college or more." Among bike commuters, ethnicities other than NH White are more likely to have lower educational attainment levels than all commuters in their race/ethnicity group. However, Hispanics that commute by bike have the highest proportion, at 78 percent, of "high school or less." This is over 21.7 percentage points higher than the share of all Hispanic commuters that have a "high school or less" education, which is by far the largest gap within race/ethnicity groups.[20] On the flip side, non-Hispanic whites that commute by bike are the only race/ethnicity group that has a higher share of those with "some college or more" than "high school or less."

Non-citizens make up over 70 percent of Hispanic bike commuters while they represent only slightly more than 36 percent of all Hispanic commuters. This is the only race/ethnicity group with a significant difference between the citizenship status of bike commuters and all commuters.

Table 5.1 Race/Ethnicity percentages of bike commuters and all commuters, New York City, 2007–2011, five-year average

	Bike commuters (%)	All commuters (%)
NH White	54.6	37.9
NH Black	6.1	21.4
NH Asian	7.8	12.9
Hispanic, any race	29.8	25.6
Other race/Ethnic group	1.6	2.2
Total	100.0	100.0

Source: 2007–2011 Five Year Average American Community Survey Public Use Micro Sample, augmented by the Center for Economic Opportunity.

Table 5.2 Educational attainment and citizenship status for bike commuters and all commuters, by race/ethnicity, New York City, 2007–2011, five-year average

| | Educational attainment (%) | | | | Citizenship status (%) | | | |
| | Bike commuters | | All commuters | | Bike commuters | | All commuters | |
	High school or less	Some college or more	High school or less	Some college or more	Citizen	Not a citizen	Citizen	Not a citizen
NH White	9.8	90.2	19.5	80.5	88.3	11.7	90.1	9.9
NH Black	46.7	53.3	41.9	58.1	83.0	17.0	82.1	17.9
NH Asian	52.2	47.8	37.3	62.7	56.7	43.3	62.4	37.6
Hispanic, any race	78.0	22.0	56.3	43.7	29.4	70.6	63.2	36.8
Other race/Ethnic group	28.6	71.4	40.0	60.0	86.8	13.2	78.6	21.4
Total	35.8	64.2	36.4	63.6	67.7	32.3	77.7	22.3

Source: 2007–2011 Five Year Average American Community Survey Public Use Micro Sample, augmented by the Center for Economic Opportunity.

To look at income, we have arranged the population by percentage of the CEO poverty threshold. For this analysis, we have divided the population between those below and above 150 percent of the CEO poverty threshold. Hispanic bike commuters are substantially more likely to be below 150 percent of the threshold than all Hispanic commuters.

The data shows that Hispanics who rely on bicycling as a means of commuting in New York are lower income, less educated, and more likely to be non-citizens relative to all Hispanic commuters. This evidence suggests the existence of a cycling demographic consistent with the "invisible cyclist" in New York City. The obvious next step would be to explore the residential location of this population relative to the bicycle infrastructure. Unfortunately, due to the small sample we have aggregated for both geographic areas and race/ethnicity groups for our spatial analysis we are unable to explore more fine-grained analyses.

Geographic analysis

New York City is comprised of 55 Census Bureau-defined Public Use Microdata Areas (PUMAs) and 15 Super-PUMAs, which are aggregates of multiple PUMAs. The Census Bureau uses these geographic areas for data tabulation and reporting. Figures 5.2 and 5.3 overlay the city's 2013 bicycle infrastructure (NYCDOT 2013b) and its Super-PUMAs.[21] Figure 5.2 shows each Super-PUMA's share of all New York City bike commuters, as well as each Super-PUMA's share of all New York City commuters. Figure 5.3 shows the percent of bike commuters who are low-income in the Super-PUMA compared to all commuters who are low-income in the Super-PUMA.

Much of New York City's bicycle infrastructure development has occurred in Manhattan and northwestern Brooklyn around the East River

Table 5.3 Distribution above and below 150 percent of CEO poverty thresholds, bike commuters and all commuters, by race/ethnicity, New York City, 2007–2011 five-year average

	Bike commuters (%)		All commuters (%)	
	Below 150%	150% and above	Below 150%	150% and above
NH White	17.7	82.3	16.3	83.7
NH Black	22.5	77.5	30.5	69.5
NH Asian	35.3	64.7	36.5	63.5
Hispanic, any race	48.2	51.8	42.2	57.8
Other race/Ethnic group	20.4	79.6	29.1	70.9
Total	27.8	72.2	28.8	71.2

Source: 2007–2011 Five Year Average American Community Survey Public Use Micro Sample, augmented by the Center for Economic Opportunity.

bridge crossings. The East River bridges provide vital connections between Brooklyn and Manhattan, enabling crucial physical connections within the bicycle network. Development of this section of infrastructure has been essential to its overall success (Pucher et al. 2010). A recent Census Bureau report indicates that Manhattan's daytime population almost doubles to over 3 million people due to the daily influx of commuters (McKenzie et al. 2013).

In Figure 5.2, the four total Super-PUMAs in lower Manhattan and northwestern Brooklyn, the area with the largest bike infrastructure, represent over half of all New York City's bike commuters while only representing 28 percent of all commuters. These areas are some of the densest in terms of bike commuters and follow overall commuting patterns. For example, as one moves further away from Manhattan, we see bike commuter shares for Super-PUMAs dropping steadily off with only 2.4 percent of all bike commuters living in eastern Queens, eastern Brooklyn, and the northern Bronx and only 1.2 percent of all bike commuters living in Staten Island.

As shown in Figure 5.3, the two Super-PUMAs in lower Manhattan have very low percentages of low-income bike commuters, at 14.3 percent and 11.9 percent. This is most likely a result of these areas having the city's lowest percentages of low-income commuters. Following this pattern, many

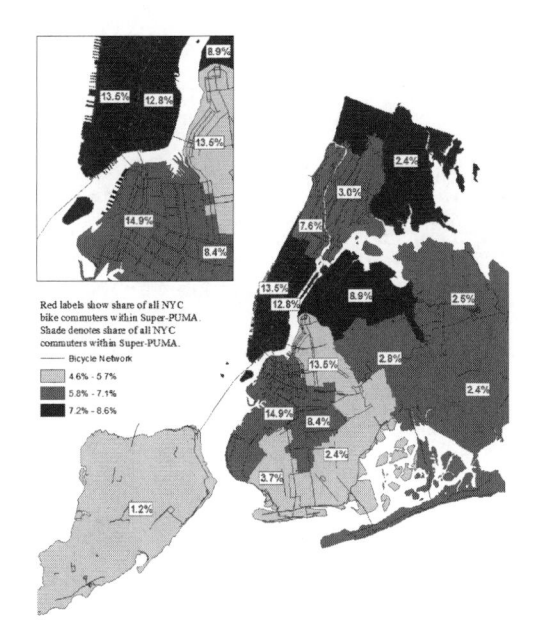

Figure 5.2 Bicycle network and share of bike commuters and all commuters with super-PUMA, New York City.

Source: New York City Department of Transportation, American Community Survey Public Use Micro Sample, augmented by the Center for Economic Opportunity, 2005–2011.

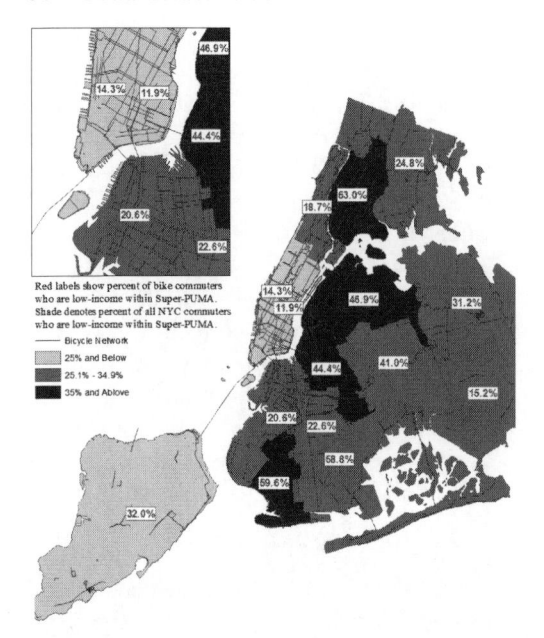

Figure 5.3 Bicycle network and percentage of bike commuters and all commuters who are low-income within super-PUMA, New York City.

Source: New York City Department of Transportation, American Community Survey Public Use Micro Sample, augmented by the Center for Economic Opportunity, 2005–2011.

of the areas with higher percentages of low-income bike commuters also have higher percentages of low-income commuters. The Super-PUMAs with the highest fractions of low-income bike commuters are located in Brooklyn, Queens, and the Bronx with bicycle infrastructures less developed than at the core. This is consistent with other report's findings that indicate that the bike infrastructure excludes large numbers of cyclists in the outer boroughs (Applebaum et al. 2011). These reports have made recommendations that attention be paid to developing the bicycle infrastructure in these parts of the city as a matter of transportation equity and social justice (Applebaum et al. 2011; Pucher et al. 2010).

How would more bike commuters affect the CEO poverty rate?

In 2011, the ACS indicated that less than 1 percent of all New Yorkers commute by bike. However, the improved bike infrastructure, coupled with New York City's climate, land use pattern, and the fact that New Yorkers make millions of trips short enough to be covered by bikes seem to indicate a considerable potential for bike use in New York (Pucher et al. 2010). In this section, we created a number of hypothetical scenarios of what would happen to the CEO poverty rate if New Yorkers did, in fact, increase their use of bikes to commute.

Here we substituted other American cities' bike commuter rates in place of our 0.8 percent. We used Portland, OR, and Seattle, WA, with 6.3 percent and 3.5 percent bike commuting rates, respectively. We also use a possibly unattainable, but non-the-less desirable, bike-commuting rate of 15 percent to analyze the economic impacts associated with a high percentage of bike commuters. To do this we created a statistical model to assign a probability value to each commuter of how likely they would be to commute via bicycle. The model is based on demographic characteristics of current New York City bike commuters. Once the probability was calculated we added the most likely bike commuters to the share that currently commute via bicycle to reach our target rates. The individual characteristics, coefficient values, and their statistical significance are provided in Table 5.4.

Table 5.4 Logit regression model of probability that commute uses a bike, 2011 American Community Survey estimate

Characteristics	Coefficient	S.E.	Significance
Constant	−4.92720	0.40682	***
Male	0.65462	0.13524	***
Age	−0.00933	0.00499	.
Works Full-Time, Year Round	−0.26138	0.13708	.
Unrelated Individual	0.48977	0.13798	***
Educational attainment (Graduate School omitted)			
Less than High School	−0.28035	0.23579	
High School	−0.58271	0.21259	**
College	−0.23532	0.17753	
Borough (Brooklyn omitted)			
Bronx	−1.55685	0.32335	***
Queens	−1.25585	0.19094	***
Manhattan	−0.54689	0.16329	***
Staten Island	−1.70334	0.45824	***
Citizenship status (Non-citizen omitted)			
Citizen by birth	−0.32010	0.16892	.
Naturalized citizen	−0.90126	0.24451	***
Race/Ethnicity (NH Black Omitted)			
NH White	1.67445	0.30736	***
Hispanic	1.67891	0.32151	***
NH Asian/Other race/ethnicity	0.55507	0.38495	

Signif. codes: 0 '***' 0.001 '**' 0.01 '*' 0.05 '.' 0.1 ' ' 1

Source: 2011 American Community Survey Public Use Micro Sample, augmented by the Center for Economic Opportunity.

In our model, every newly assigned bike commuter saves what they would have originally spent on commuting cost. This translates to more money in the commuter's budget to spend on items in the threshold such as food, clothing, and shelter. Table 5.5 shows the three different scenarios along with our baseline New York City estimates. In all scenarios commuters save between $750 and $1,000 on average annually, when they switch to bike commuting. Obviously, as we assign more bike commuters, the aggregate dollar value saved in commuting costs increases. In the Portland target scenario, commuters save over $95 million while the aggregate value tops out at over $538 million in the 15 percent scenario.

There are limited effects on the CEO poverty rate in our hypothetical scenarios; this is unsurprising. The percentage point decreases in the CEO poverty rate for the whole population is very small in the Seattle hypothetical which moves 1,318 people out of poverty. Almost 15,000 people are moved from poverty using the 15 percent bike-commuting rate hypothetical, which reduces the CEO poverty rate by 0.2 percentage points. However, these marginal changes illustrate the proficiency of the CEO Poverty Measure to assess the tangible financial impacts for families of potential policy changes and the aggregate overall effect on the CEO poverty rate.

One drawback in this model is that we are assuming that the demographic and economic characteristics of hypothetical bike commuters will be similar

Table 5.5 New York City CEO poverty rates and hypothetical CEO poverty rates, New York City, 2011 American Community Survey

	New York City currently	*Seattle hypo*	*Portland hypo*	*15% hypo*
Bike commuter percent of all commuters	0.8	3.5	6.3	15.0
Commuting costs saved				
Mean per commuter		$747	$867	$985
Median		$630	$840	$1,050
Sum		$95.3M	$199.1M	$538.5M
New York CEO poverty rate (%)	21.3	21.2	21.2	21.1
New Yorkers moved out of poverty		1,318	7,823	14,589

Source: 2011 American Community Survey Public Use Micro Sample, augmented by the Center for Economic Opportunity.

to those that currently commute by bike. If it was possible to target the population that would move from more expensive commuting methods to bicycling, especially those who are poor and those who are vulnerable to poverty, we could have a larger impact on the CEO poverty rate.

Discussion

There are a number of issues that can be addressed when looking at the next steps for researching the economic impact of urban bicycle commuting. The first part of this section looks at the methodology of this chapter and explores methodological issues that could be addressed. The second section explores the issues associated with commuting costs and the CEO Poverty Measure.

One issue that needs to be addressed when doing further work on this subject is the statistical model used in the hypothetical scenarios. The model was created using the 2011 ACS sample. In future work, we plan to use a combined five-year dataset to address the small sample size issue. Also, New York City's fraction of bike commuters is very small. For every bike commuter with a certain set of characteristics, there are many with similar characteristics that use alternative commuting methods, making the probability values of bike commuting for our model very small. It would be appropriate to examine this model further to explore alternative methods.

Also, as we mentioned earlier, this model assumes that the demographic and economic characteristics of hypothetical bicycle commuters would be similar to those of current bicycle commuters. We cannot actually assume that this will be the case. For example, currently less than a quarter of all New York City bike commuters are women. Research has shown that women are more likely to use bikes when they feel safe (Szczepanski 2013). With an expanding bicycle infrastructure that makes biking both easier and safer, it is not unreasonable to assume that more women would use bikes to commute to work.

As mentioned earlier, we use a $0 per trip bicycle commuting cost in our model. During the research for this paper, a number of resources indicated that the cost was between $240 to $300 annually for bike maintenance, bike parts, and wear-and-tear. In future iterations, we will explore altering the annual cost of commuting via bicycle to reflect those costs. These changes will have an effect on the budgets of those that commute via bicycle.

Conclusion

New York City's bicycle infrastructure is expanding. The NYDOT has a goal of 1,800 miles of bike paths, lanes, and routes by 2030 (Pucher et al. 2010). Data suggest that there is a current and future low-income bicycling constituency that could benefit from this expansion, an incredible opportunity to positively impact low-income families' budgets. There are positive signs that officials are aware of this possibility. Citi Bike is offering discounted annual memberships to residents of the New York City Housing Authority (Citi Bike NYC 2013a).

In Los Angeles, programs such as City of Lights run by the Los Angeles County Bicycle Coalition are aimed at providing low-income Hispanic immigrants with bicycle lights and safety information with the intention that they can educate and advocate in their communities. In New York, Recycle-a-Bicycle's youth program teaches basic bike mechanics to volunteer youth after school in exchange for a bicycle frame. Similar public programs that foster cycling advocacy among the low-income population while lowering the economic and social barriers to bicycling could be an important component of New York City's bicycle future.

Attention should also be paid to the current and future bicycle infrastructure. Our data shows that the neighborhoods with the largest percentages of low-income bike commuters are areas with limited bicycle infrastructure in the outer boroughs. While these areas also have relatively small shares of the city's overall bicycle commuters, Pucher et al. have argued that expansion of the bike network in these neighborhoods is a matter of transportation equity and social justice (2010). The 'invisible' nature of this low-income bike commuting population suggest that further research be done to ensure that any future expansion of the bicycle infrastructure benefits all bicyclists throughout the city.

Notes

1. In 2017 the CEO Poverty Measure was renamed the NYCgov Poverty Measure and the Center for Economic Opportunity became the Mayor's Office for Economic Opportunity.
2. Citi Bike will expand to 12,000 bikes and 750 stations by end of 2017 (NYCDOT 2016).
3. It is important to note that the American Community Survey only collects journey to work data, therefore other leisure and non-work bike trips are not recorded.
4. This concern turned out to be unfounded under Mayor de Blasio's Vision Zero program, New York City projected 18 miles of additional protected bike lanes and 75 additional unprotected bike lanes by end of 2016 (Del Valle 2006).
5. New York City was the first municipality within the United States to adopt an alternative poverty measure. The CEO Poverty Measure methodology follows recommendations made by the National Academy of Sciences in 1995 at the behest of Congress. The Supplemental Poverty Measure, adopted by the Obama Administration, uses a similar methodology to the one pioneered by New York City.
6. The Food Stamp program was recently renamed the Supplemental Nutritional Assistance Program (SNAP). Since the program is more widely recognized by its former name, we continue to use it.
7. CEO's first report followed the NAS methodology. In 2011, the Census Bureau released a Supplemental Poverty Measure (SPM). The SPM had slight methodological differences from the NAS recommendations. To insure comparability with the SPM, CEO adopted much of the SPM methodology. These differences

are discussed in Appendix B of our latest report, "The CEO Poverty Measure, 2005–2011" which can be viewed here: www.nyc.gov/html/ceo/downloads/pdf/ceo_poverty_measure_2005_2011.pdf.

8. The NAS reference families are those composed of two adults and two children. The threshold for this family is then scaled for families of different sizes and compositions. See Appendix B of "The CEO Poverty Measure, 2005–2011."

9. The NAS suggested that this point lay between the 30th and 35th percentile. (Citro and Michael, 1995, 106.)

10. The SPM has three different thresholds, one for renters, one for homeowners with a mortgage and one for homeowners without a mortgage. The rationale for this is discussed in Appendix B of our latest report.

11. The CEO Poverty Measure accounts for inter-area costs by adjusting the housing portion of the threshold. We use a ratio of the U.S. Housing and Urban Development's (HUD) Fair Market Rent (FMR) for New York City versus the U.S.-wide FMR.

12. http://www1.nyc.gov/site/opportunity/poverty-in-nyc/poverty-measure.page

13. We round to the nearest whole number for the number of workdays.

14. In 2008, the WKW variable was changed from the actual number of weeks to a range format. For our 2008 through 2011 calculations, we used the midpoint of each range in our calculations. We cap the number of weeks worked at 50 to account for sickness or vacation.

15. www.bikeleague.org/programs/bikemonth/pdf/BTWW_Booklet.pdf.

16. http://blumenauer.house.gov/index.php?option=com_content&task=view&id=817&Itemid=167.

17. www.nyc.gov/html/dot/downloads/pdf/cycling-in-the-city.pdf.

18. The ACS only records journey to work trips therefore other non-commuting bike trips are not recorded. It also asks respondents for their usual method of commuting. Therefore if someone only bikes twice a week and commutes via subway the other three days, they would respond that they take the subway therefore undercounting the bicycle commuting numbers. www.nyc.gov/html/dot/html/bicyclists/ridership-facts.shtml

19. For example, in 2011 the American Community Survey indicates that 30,459 New York City residents commuted via bike, which is an un-weighted sample of 253.

20. Non-Hispanic Asians have the next highest difference between those that commute by bike and all commuters at 15 percentage points. An important next step would be to look at the occupation for non-Hispanic Asians and Hispanics to examine their occupations and industry.

21. PUMA and Super-PUMAs are statistical geographic areas defined, by the Census Bureau, for the tabulation and dissemination of survey data. www.census.gov/geo/reference/puma.html.

References

Applebaum, Max, Andrew Camp, Conor Clarke, Joe Delia, Jennifer Harris-Hernandez, Sungbae Park, Brian Paul, Scott Richmond, Sam Stein, Matthew Wallach, and Sung Hoon Yoo. 2011. *Beyond the Backlash: Equity and Participation in Bicycle Planning*. Hunter College Department of Urban Affairs and Planning. www.streetsblog.org/wp-content/uploads/2011/05/BeyondBacklash2011.pdf.

Citi Bike NYC. 2013. *Citi Bike Blog*. http://citibikenyc.com/blog.

Citi Bike NYC. 2013a. *Discounted Annual Memberships*. www.nyc.gov/html/dot/html/about/datafeeds.shtml.

Citro, Constance F. and Robert T. Michael (Eds.). 1995. *Measuring Poverty: A New Approach*. Washington, D.C.: National Academy Press.

Davis, Paul. 2011. "Are Bike Lanes Expressways to Gentrification?" *Shareable.net*. www.shareable.net/blog/are-bike-lanes-an-expressway-to-gentrification.

Del Valle, Gaby. 2006. *NYC will Build 18 Miles of Protected Bike Lanes in 2016*. Gothamist.com. http://gothamist.com/2016/09/13/more_bike_lanes_yay.php?utm_source=feedly&utm_medium=webfeeds.

Flegenheimer, Matt. 2013. "Bike-Share System for New York is Built with Ideas from Around the World." *New York Times*. www.nytimes.com/2013/05/22/nyregion/a-bike-share-system-for-new-york-built-from-ideas-around-the-world.html?_r=1&.

Fuller, Omari and Edgar Beltran. 2010. "The Invisible Cyclists of Los Angeles." Plannersnetwork.com. www.plannersnetwork.org/2010/07/the-invisible-cyclists-of-los-angeles/.

Huff, Kristin. 2011. *Low-income Latino Cyclists in Los Angeles County: A Socio-economic Spatial Analysis*. http://lacbc.files.wordpress.com/2010/06/huff_mccormick_up259_final_report.pdf.

Leung, Adrian and Allison Mannos. 2011. *Bicycling is for Everyone: The Connections Between Cycling in Developing Countries and Low-Income Cyclist of Color in the US*. Streetsblog LA. http://la.streetsblog.org/2011/06/01/bicycling-is-for-everyone-the-connections-between-cycling-in-developing-countries-and-low-income-cyclists-of-color-in-the-u-s/.

Levitan, Mark, Christine D'Onofrio, John Krampner, Daniel Scheer, and Todd Seidel. 2008. *The CEO Poverty Measure: A Working Paper by the New York City Center for Economic Opportunity*. www.nyc.gov/html/ceo/downloads/pdf/final_poverty_report.pdf.

Levitan, Mark, Christine D'Onofrio, John Krampner, Daniel Scheer, and Todd Seidel. 2013. *The CEO Poverty Measure: A Working Paper by the New York City Center for Economic Opportunity*. www.nyc.gov/html/ceo/downloads/pdf/ceo_poverty_measure_2005_2011.pdf.

McKenzie, Brian, William Koerber, Alison Fields, Megan Benetsky, and Melanie Rapino. 2013. *Commuter- Adjusted Population Estimates: ACS 2006–2010*. Washington, D.C.: U.S. Census Bureau. www.census.gov/hhes/commuting/files/ACS/Commuter%20Adjusted%20Population%20Paper.pdf.

Mineta Transportation Institute. 2011. *Getting Around When You're Just Getting By: The Travel Behavior and Transportation Expenditures of Low-Income Adults*. http://transweb.sjsu.edu/MTIportal/research/publications/documents/2806_10--02.pdf.

NYCDOT. 2009. *NYC DOT and DOB Announce Bikes in Buildings Law to Take Effect*. www.nyc.gov/html/dot/html/pr2009/pr09_052.shtml.

NYCDOT. 2012. *PlaNYC Bicycle Network Expansion: New York City Department of Transportation*. www.nyc.gov/html/dot/downloads/pdf/bikeroutedetailsfy07-fy12.pdf.

NYCDOT. 2013. *Bicycling, NYC: New York City Department of Transportation*. www.nyc.gov/html/dot/html/bicyclists/bicyclists.shtml.

NYCDOT. 2013a. *Bicycle Counts: New York City Department of Transportation*. www.nyc.gov/html/dot/html/bicyclists/bike-counts.shtml.

NYCDOT. 2013b. *Cycling in the City: An Update on NYC Cycling Counts*. www.nyc.gov/html/dot/downloads/pdf/2013-nyc-cycling-in-the-city.pdf.

NYCDOT. 2016. *Manhattan Community Board 8 Transportation Committee.* http://a841-tfpweb.nyc.gov/bikeshare/files/2016/07/Manhattan-CB-8-Updated-Final-Plan-Presentation.pdf.

Pucher, John, Lewis Thorwaldson, Ralph Buehler, and Nicholas Klein. 2010. "Cycling in New York: Innovative Policies at the Urban Frontier," *World Transport Policy and Practice*, Vol. 16, Summer 2010. http://policy.rutgers.edu/faculty/pucher/cyclingny.pdf.

Szczepanski, Carolyn. 2013. "Building a Woman Bike Friendly America: Two of the nation's top bike researchers weigh in on how to get more women riding." *American Bicyclist.* www.bikeleague.org/members/pdfs/March-April2013-forweb.pdf.

Tavernise, Sabrina. 2011. "A Population Changes, Uneasily." *New York Times.* www.nytimes.com/2011/07/18/us/18dc.html.

Victoria Transport Policy Institute. 2011. *Transportation Cost and Benefit Analysis II—Vehicle Costs.* www.vtpi.org/tca/tca0501.pdf.

Wall Street Journal. 2013. "Dorothy Rabinowitz Opinion: Death by Bicycle." http://live.wsj.com/video/opinion-death-by-bicycle-/C6D8BBCE-B405-4D3C-A381-4CA50BDD8D4D.html?KEYWORDS=rabinowitz#!C6D8BBCE-B405-4D3C-A381-4CA50BDD8D4D.

6 Middle modalism

The proliferation of e-bikes and implications for planning and urban design

Derek Chisholm and Justin Healy

Introduction

Motorized bicycles are different from non-motorized bicycles. They have greater range, cost, and weight as well as higher average and top speeds. For this reason, powered bikes, a type of Light Electric Vehicle (LEV), require consideration by urban planners and designers. Key differences for urban planners include presumed larger bike-sheds—the potential capture area for trips around any given destination—and higher speeds, which will affect the design of shared "bike" facilities. As of yet, the planning community has not developed specialized treatments and approaches for LEVs.

Traditionally, planners consider five major travel and commute options or modes, including walking, biking, driving an automobile, riding a bus, and using rail-based systems. Planners and designers have been working toward and encouraging multi-modalism, in which multiple modes of travel are supported and provided for within public rights of way. This allows for greater choice and opportunity for travelers. But with the proliferation of e-bikes, we must also now consider "middle-modalism." Without further refining contemporary transportation planning and urban design approaches, the modes in the "middle," LEVs, may be poorly accommodated. Since LEVs are a rapidly growing choice for many commuters, this is problematic.

This chapter focuses on e-bikes, and does not address internal combustion engine bikes. Neither does this chapter address the full array of LEVs, for example hover boards, modified golf carts, and so on. Unlike these other vehicle types, e-bike use on public roads is rapidly increasing, which requires some regulatory and design planning for successful integration into the built environment. Commuting options can be improved by accepting e-bikes as a "middle mode" and ensuring that their use is properly planned for. This chapter draws from secondary sources as well as the authors' own work and experience to explore these issues, including an in-depth literature review, an assessment of bike facility designs, the results of a modeled approach to e-bike shed planning, and the perspective of the lead author, who builds e-bikes and has commuted by e-bike since 2010.

Design vehicles

For the purposes of this study, two design vehicles were used. These two LEV types were considered in the assessment of current bike facility design standards and in the e-bike shed modeling. In brief, the two types include: e-assist bikes, commercially available bikes with speed governors—roughly 250-watt motors; and e-power bikes, which are more frequently home-built, have longer ranges, and often have higher average and top speeds. E-power bikes often have from 750 to 1,000 watts of power. Both types of bikes can be throttle or pedal assisted, though often the slower e-assist bikes require the rider to pedal to activate the motor.

Physically, there is little difference between the two e-bikes; they have similar sizes and features. The builders of the often-faster home-built/DIY bikes, though, will sometimes choose heavier and larger frames and wheels for their bikes. For the e-assist bikes types, the batteries can be sealed lead acid (SLA/PbA), nickel metal hydride (NiMH), nickel cadmium (NiCad), or one of the many lithium polymer (e.g. LIFEPO) packs of cells. The less expensive, non-lithium batteries have lower power to weight ratios. For the faster, e-power bikes, the energy demands almost always require lithium batteries. Otherwise the vehicle would have to haul up to 50 pounds of SLA batteries to attain the needed watts. So the e-power bikes require more stored energy and heavier motors.

It is necessary to differentiate between these two types of LEVs, as it is also necessary to differentiate e-bikes from bikes and electric cars. In the United States, automotive companies have worked to dominate the market-place for individual vehicular travel. As well, the nation's investment in auto-only interstate and roadway systems has contributed to the singularity in vehicular travel options over the last 70 years. But now, public policy, innovation, and the deficiencies of auto-centric systems are driving the proliferation of a variety of LEVs, including e-bikes.

If our methods and approaches to urban design and planning do not evolve concurrently with these innovations, these will be misunderstood and poorly accommodated travel modes. Instead of planning for only three speeds and types of travel—walk, bike, drive—it is now important to develop more sophisticated design treatments and policy, accommodating a wider range of community and recreational travel options including jogging, skating, cycling, e-cycling, riding a Segway, using people movers, or participating in car sharing programs.

The e-assist bike performs much like a conventional bike. The majority of the early models available in the United States came from Europe, and had built-in governors limiting motor-driven speeds to 20 miles per hour. Many conventional bikes are pedaled at this speed for short distances during a typical work commute. The e-assist bikes offer more range and, perhaps, a higher average speed than regular, non-powered bikes. Though, at "normal" speeds, the time required for the commute can also become a constraint and a determinant of mode choice. E-assist bikes have a greater range and will provide a larger travel shed compared to a non-powered bike. But for the design of roadway facilities, these vehicles can be grouped with all other bikes, because they will travel in platoons at the speed of other bikes, and generally fit into existing systems. One exception is in bike parking. Because e-bikes cost from $1,000 to $5,000, there is a need for greater numbers of fully enclosed lockers, ensuring the security of these investments. Once the appropriate bike facilities are provided, more riders may choose to purchase an e-bike with the confidence of knowing that it will not just be locked to a rack on the sidewalk.

Han Goes, a product designer and developer, has asserted that the e-bike market will evolve beyond simply affixing motors and batteries to conventional bikes, and will lead more toward a new vehicle type or mode (Goes 2009). He explains that current practice focuses on 'electrifying' existing bicycle concepts, but that the industry should and will come to a more sophisticated understanding of how to design a vehicle that is not a bike, and not a car, but is tailored—regardless of categorical presumptions about the typical three vehicle types (walk, bike, drive)—to the consumer's needs. Goes refers to our current practice as *horizontal product differentiation*, applying the bike concept to different trip and rider needs, as opposed to *vertical product differentiation*, that would seek to achieve the goals of LEVs and develop different design concepts for such.

There are now some fast, commercially available e-bikes. Many of these, such as the dual-motor Stealth e-bike are intended and sold for recreational off-road use. A recently released street bike from the e-bike company, A2B (Figure 6.1) is capable of a higher-than-usual top speed of 28 miles per hour. Its 500-watt motor and battery provide a range of up to 37.5 miles before the bike needs to be charged again.

Electric bicycle regulations

In the United States, federal law allows an e-bike to be regulated as a bicycle when it uses less than 750 watts of power, has functional pedals, and a maximum speed of less than 20 miles per hour (15 U.S.C. 2051 et seq. LOW-SPEED ELECTRIC BICYCLES SEC. 38. (a)). Commercial e-bikes exceeding these limits are regulated as motor vehicles. They cannot be ridden and parked on sidewalks or in bike lanes, and must meet additional requirements. These limits do not apply to home-built e-bikes, which are regulated by state law. Both Oregon and Washington have similar but different limits from the U.S. Department of Transportation; they require that e-bikes conform to less than 20 mph and less than 1,000 watts of power.

According to the Revised Code of the State of Washington:

> "Electric-assisted bicycle" means a bicycle with two or three wheels, a saddle, fully operational pedals for human propulsion, and an electric

Figure 6.1 Unveiling the new A2B in New York.
Source: Electric Bike Report.

motor. The electric-assisted bicycle's electric motor must have a power output of no more than 1,000 watts, be incapable of propelling the device at a speed of more than twenty miles per hour on level ground, and be incapable of further increasing the speed of the device when human power alone is used to propel the device beyond twenty miles per hour.

(WA RCW 46.04.169)

Proliferation: demand and constraints

E-bikes are in demand across the United States, Europe, and Asia. Navigant Research, a market research and consulting group, forecasted in early 2012 that the total 2012 sales of e-bikes would surpass 30 million, making e-bikes the world's best-selling electric vehicle. They projected that global sales would increase at a compound annual growth rate of 7.5 percent between 2012 and 2018 (Martin 2012). At the 2012 National Interbike Show in California, an annual bicycle exposition for retailers, media, importers, and distributers, the number of exhibits featuring e-bikes more than doubled (Meyerson 2013). The Evelo Company is so convinced of the ability of e-bikes to win over consumers that, in 2013, they provided ones for a free,

30-day period in eight cities, including Redmond and Seattle, Washington. Participants exchanged their car keys for an e-bike and were required to use only the e-bike, public transit, or car-sharing for 30 days (Evelo 2013).

In China

"For all the talk of China's growing infatuation with automobiles, the world's most populous nation continues to roll primarily on two wheels—and, increasingly, an electric motor drives them" (Fairley 2005). In China there are 450 million bikes and 100 million e-bikes (Weinert 2007). The vast majority of these e-bikes were produced after 2005. The China Bicycle Association, a bicycle industry group in China, collects data on e-bike and bike usage and sales, and has found greatly increasing sales and production in China. Even a decade ago, nearly 20 million electric bikes were sold in China annually, which is a rate almost twice the 10.9 million new bikes sold in the United States in 2008 (Cherry, Weinert, and Xinmiao 2009).

The Castellan-AG industry group, located in Germany, also collects data on Chinese e-bike production, as well as in Europe and the United States. They typically share research about the appeal and use of e-bikes through the range of urban morphological transects. Biking is generally the focus of urban and small town planning in the United States. In more rural areas in the U.S., there is less focus on biking facilities for commuting, though the provision of multi-use paths and trails for recreational purposes is widespread and was heavily invested in through the Safe, Accountable, Flexible, Efficient Transportation Equity Act (SAFETEA-LU) Enhancements Program. Assuming the same primarily urban application of e-bikes is another result of categorically lumping together bikes and e-bikes, the greater range, speed, and perhaps rider confidence allows the e-bike to affect mode-share in populations and geographies that have gone largely unaffected by other attempts to promote cycling. Table 6.1 shows remarkably similar receptiveness among the Chinese, from villages to large cities.

Table 6.1 Reported interest in e-biking in China by urban form

Urban Form	Reported interest in e-biking				
	Certainly	Maybe	Probably not	Definitely not	Already own e-bike
Village	27	30	23	19	< 1
Small City	23	36	24	17	0
City	25	31	23	21	0
Large City	21	30	23	26	0
Total	25	31	23	21	0

Source: Castellan-AG 2012.

In Europe

Demand for e-bikes is growing in Europe, where Germany and the Netherlands are the largest markets and, in France, Paris has a program to provide subsidies to electric bike purchasers (Galbraith 2012). A report focused on the Dutch e-bike market found that, while only 3 percent of the Dutch population owned an e-bike, 40 percent of the Dutch people were interested in e-biking (Hendriksen et al. 2008). The study included data on why people were interested in owning an e-bike, and the responses are listed here in descending order of popularity based on frequency of positive responses (Hendriksen et al. 2008, 24):

1.   to make cycling with headwind easier;
2.   to be able to cycle over longer distances without (much) extra effort;
3.   to make it easier to climb hills;
4.   I am not very sporty but I would like to have (some) more exercise;
5.   to cycle faster (less travel time) without (much) extra effort;
6.   as an alternative to less environmentally friendly means of transport;
7.   to get to work without sweating.

In Europe, and especially in the already bike-oriented multi-modal urban areas, e-bikes have greatly increased in number and continue to do so. Many governments at the national and city scale recognize the benefits of proliferating LEVs and have both policy and investment initiatives to encourage the mode shift.

In the United States

In the U.S., e-bike stores have opened in many cities including Austin, TX; Chattanooga, TN; Chicago, IL; Houston, TX; Portland, OR; San Diego, CA; Seattle, WA, and more. There are electric bike-sharing projects in San Francisco, California; Washington, D.C.; and Chattanooga, TN. More than 100,000 e-bikes were planned to be sold in the U.S. in 2012 with sales projected to more than triple by 2018 (Galbraith 2012).

Constraints: rider needs and priorities

The City of Portland, OR has won numerous awards for its bike friendly plans and design; and public and private planners in the city have started to study and consider LEVs. But other than John MacDonald's and Jennifer Dill's work with The Oregon Transportation Research and Education Consortium (OTREC), little analysis has been completed for e-bikes and other LEVs. In February 2010, the City of Portland approved the Portland Bicycle Plan for 2030. It is an inspired and ambitious plan, addressing land use, transportation, planning, design, and behavioral issues. Central to the land use planning elements is the concept of a "20-minute neighborhood." In a 20-minute neighborhood, "residents live within a short walk or bicycle ride to daily destinations." The map shown in Figure 6.2 is from Portland's analysis, and depicts the accessibility of neighborhoods with pale gray being the most accessible, and dark gray the least.

The Portland Bicycle Plan does not define or examine the *bicycle* vehicle type. The city's designation of a 20-minute neighborhood is based on certain assumptions about the bike shed. A 20-minute neighborhood bike shed is not just a simple circle based on speed and distance. It is determined by numerous factors, including: average bike speed; distance; the presence of steep hills; the speed of cars; safety and perceived safety of intersections; number of intersections; crime and perceived crime; presence of parked cars; presence of bike lanes, green boxes, and so on; adequacy of bike storage at destinations; roadway maintenance; and physical condition of the rider. Of these 12 factors, e-bikes perform differently in eight categories, and significantly different in some of these (see Table 6.2). The degree to which the e-bike performance differs from non-motorized bikes is dependent on the e-bike itself. The difference between e-bikes and bikes, as shown here, is the reason why further research, planning, policy, and infrastructure questions should be further explored.

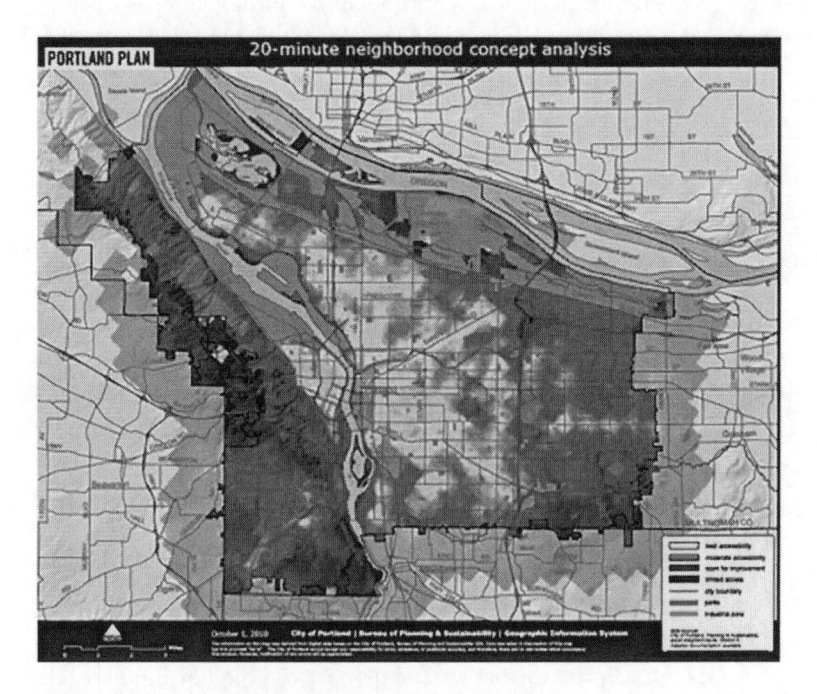

Figure 6.2 20-minute bike shed.

Source: City of Portland, Department of Transportation.

Table 6.2 City of Portland 12 factors for bike shed planning, expanded to include e-biking

Average bike speed	E-bikes have higher average speeds, higher average low speeds, and sometimes (for *e-power* bikes) higher top speeds.
Distance	Recreational cyclists can ride very long distances. Frequently groups of friends bike one, two, or more marathon distances on a sunny weekend. But non-powered commute trips are constrained by time and the adverse effects of generating energy from food and muscles—perspiration. Many commuters are biking one to four miles to work. Few pedal for eight or more. E-bikes have the ability to complete a work-based commute, in a reasonable amount of time, without requiring the rider to change clothes and shower at their destination.
The presence of steep hills	The ability to more easily climb hills with an e-bike is one of the strongest reasons for purchasing one.
The speed of cars	E-bikes, especially the faster e-power bikes, are able to "take a lane" in most city traffic where posted speeds are 25 or 30 mph. To a small extent, and for confident riders, an e-bike may allow the rider more comfort in riding among autos, at their speed, with the flow of traffic.
Safety and perceived safety of intersections	To a small degree, e-bikes increase one's confidence in busy intersections. Non-powered bikes can attain high speeds, but are not easy to rapidly accelerate from a full stop. With pedaling alone, overcoming inertia requires more power from the rider. If, for example, a rider is waiting at the minor movement at a two-way, stop-controlled intersection with an arterial roadway, and is waiting for a safe "gap" in the traffic, the quick acceleration of an e-bike aids the rider in crossing or merging.
Number of intersections	Frequent intersections increase connectivity but also increase conflict points among the vehicles and pedestrians. E-bikes perform like other bikes in regards to this factor.
Crime and perceived crime	Perhaps quicker acceleration could make a rider feel safer in an otherwise, unsafe area. But the increased value of the vehicle could cause the rider to worry.
Presence of parked cars	E-bikes perform like other bikes in regards to this factor.
Presence of bike lane, green boxes, etc.	Though quicker acceleration may give a rider more confidence, and less need for the green boxes, this is speculative and is not considered a factor in this study.
Adequacy of bike storage at destinations	As described in this study, the higher costs of e-bikes increases the need for safe storage at destinations.
Maintenance of the roadway	E-bikes perform much like other bikes in regards to this factor. E-bikes are often heavier and may not be as manageable through gravel, especially if the battery weight is born high on the vehicles. Seldom, however, do e-bikes have very narrow road tires (<1.75in) because of the heavy components.
Physical condition of the rider	Another strong reason for increased e-bike sales is that they enable a number of non-athletes, senior citizens, and disabled riders to enjoy longer more frequent trips by bike. This factor does not affect the e-bike travelshed analysis. But it does affect the mode share for bikes, providing a new opportunity for many individuals for whom biking was not a desirable option (Dill and Rose 2012). In the Netherlands, nearly one million e-bikes are in use, mostly by elderly people (Parker 1999).

Source: Portland Bureau of Transportation 2010, augmented by authors.

Rider types

Roger Geller, Bicycle Coordinator for the City of Portland, has defined four types of cyclists. He then analyzed these four groups, the policy and program initiatives most suitable for each group, as well as additional implications for planning and design. These groups are "strong and fearless," "enthused and confident," "interested but concerned," and "no way no how" (Geller 2006). See Table 6.3.

How would the proliferation and availability of e-bikes change behavior among these four groups? Additional research will be necessary to determine such, but it is reasonable to believe that residents currently among the no way no how group, and whose primary impediment is a steep hill or a bad knee, may start to ride in increasing numbers. Also, those in the two middle groups, especially the enthused and confident, may choose to switch from other modes and pedal the longer commutes (between 5 and 15 miles) on an e-bike. Lastly, members of the two middle groups may also choose to ride more often to work, knowing that the electric motor can keep them from perspiring heavily.

E-bike shed analysis

According to the National Household Travel Survey (NHTS), biking and walking make up 11.9 percent of all trips made in this country (NHTS 2010). Riding a bicycle is more energy efficient than walking and enables

Table 6.3 Geller's four types of cyclists, augmented by authors

Type	Percentage of Portland's population	
Strong and fearless	< 0.5%	
Enthused and confident	7%	Increased riding by this group is the primary reason that bike commuting in Portland doubled between 1990 and 2000, driven partly by investments in bike-friendly infrastructure.
Interested but concerned	60%	
No way no how	35%	Geller states that this group is prevented from riding by factors that are very difficult to address with programs and infrastructure. But there are factors among these, such as topography and physical inability that *could be addressed by e-bikes.*

Source: Geller 2006.

the average rider to go 3.5 times as far as a pedestrian (NHTS 2010). Table 6.4 (below) shows mode choice for short trips.

According to the survey data, 59 percent of all bike trips are one mile or less and 85 percent are three miles or less (NHTS 2010). This speaks to issues of range for most bike commuters, and the ability of e-bikes to greatly change current biking trends.

The term "travel shed" is used to describe the area accessible in a typical commute by different modes. Bike sheds have been found to include up to ten times as many homes in comparison to the same destination's walk shed; and the e-bike increases the number of accessible homes by a factor of 30 over walking (Parker 1999). E-bikes promise increased travel sheds with minimal need for new infrastructure or other investments. Previous e-bike travel shed research by Alan Parker yielded the results shown in Table 6.5.

In another study, survey participants indicated they would not shift from driving to biking if they have to travel more than 7 kilometers (4.35 miles) (Hendriksen et al. 2008). As part of an OTREC project, Dill and Rose studied 21 blogs, mining them for discussions of e-bikes, and found that e-bikes were successfully supplanting automobile trips on journeys of up to 24 km (15 miles). They also found that e-bikes mitigated the adverse effects of steep grades for many commuters (Dill and Rose 2012, 5).

Justin Healy, owner of Real Urban Geographics in Portland, completed the following modeled analysis of e-bike sheds. The analysis was originally conducted as part of the Oregon Department of Transportation (ODOT)

Table 6.4 Mode choice for short trips

	<1 mile	*<2 mile*	*<3 mile*
% of all trips of any length	28	40	50
Mode			
Bike (%)	2.3	2.0	1.8
Walk (%)	35	26	21
Car (%)	60	68	72

Source: National Household Transportation Survey.

Table 6.5 Travel shed comparison

Factors	*Mode*			
	Walk	*Bike*	*Racing bike*	*E-bike*
Speed (km/h)	6.1	20	25	24
Distance (km)	0.8	2.5	3.2	7
Catchment area (sq. km)	1.3	12.4	20	40

Source: Parker 1999.

project for the Southwest Corridor Plan in the Portland metropolitan region (Year). The goal of the analysis was to define time shed boundaries for bus, bicycle, and pedestrian facility service in the "SW Corridor" project area where high capacity transit stations and stations areas may later exist. The Portland Metropolitan Regional Government, ODOT, and TriMet, Portland's regional transit agency, identified the locations as "focus areas." These focus areas were used to test how far a commuter could travel in different time increments via walking, biking, and transit.

The methodology for the development of travel/time sheds includes several steps. First, define trip generators. In this case, ten trip generators were selected from libraries, downtown centers, parks, colleges and universities, shopping centers, and other features in the study area. Second, define valid routes from METRO street data: this method assumes that bicycles can travel on local streets, trails, and the major facilities in the corridor that have bike lanes or bike boulevards. Third, join multiple valid route data sets (streets, trails, and bicycle facilities) into a single dataset, allowing for topological connections where data sets intersect. Fourth, convert valid routes to routing network. Using ArcGIS Network Analyst, a network data set was created upon which to base travel time models. Fifth, insert barriers to travel where the network encounters pedestrian-only access or blocked access. Sixth, modify grade-separated crossings with the correct topology. Seventh, assign slope factors to each network segment and use DKS assumptions about travel speeds and slope encumbrances to calculate the travel time of a bicycle along each segment. Eighth, establish assumptions for bicycle speed (see Table 6.6). Ninth, establish time horizons for modeling travel time contours. Travel time contours were originally modeled at 15, 30, 45, and 60-minute increments. Later, when the e-bikes were modeled, a five-minute travel time contour was shown. Tenth, set traffic signal delays. In this case, delays at traffic signals were fixed at 25 seconds.

In Figure 6.3, the map represents the modeled time/travel shed for bikes, e-assist bikes, and e-power bikes. The contours display the distances that could be travelled, given the methodological assumptions above, in only five minutes. Clearly the reach of the e-bikes is evident. When measured in total acreage accessed within five minutes, the three vehicle types performed as shown in Table 6.7. The scenarios refer to the areas that were studied.

Table 6.6 Bicycle speed assumptions

	uphill (mph)	flat grade (mph)	downhill (mph)
Bike	8	10	12
E-assist bike	10	20	24
E-power bike	12	26	32

Source: Authors.

Table 6.7 Comparative modeled time/travel shed for bikes, e-assist bikes, and e-power bikes

Scenario	Bike acres	E-assist bike acres	E-power bike acres
Suburban neighborhood	1527.8	3043.7	5125.7
Urban westside	828.7	1754.7	3014.8
Urban eastside	1485.3	3237.7	5368.5
Commercial mix	756.3	1557.6	2819.5

Figure 6.3 Five-minute bike, e-assist bike, and e-power bike sheds for Portland.
Source: Urban Eastside.

The scenarios tested bike travel in different urban transects, allowing for a comparison between suburban and urban areas. The original modeling was also useful in identifying barriers to bike travel.

Speeds

Research has found that the average speed of an electric bike is 24 kmh (14.9 mph), compared with 17 kmh (10.5 mph) on a traditional bike (Hendriksen et al. 2008). Home built models are capable of cruising speeds above 20 mph and top speeds above 30 mph. Vehicles capable of such speeds can mix poorly with slower cyclists, walkers, joggers, and persons in wheelchairs or pushing baby strollers. Cyclists have expressed concern about the mixing of bikes and e-bikes because of the higher speeds of e-bikes (Dill and Rose 2012).

Hauling ability

One significant limitation of two-wheeled transport is limited hauling ability. Recent trends have shown that, with the right design and accessories, heavy loads can be managed by bike. And, of course, a little assistance with the pedaling is a natural, complementary improvement to cargo bikes and trikes. In other countries, bikes and trikes have been widely used for cargo hauling purposes (see Figures 6.4a, 6.4b).

In May of 2013, Cathy Lau from OHM Electric Bikes reported on the Electric Bike Report blog that the Other Coast Café in the Capitol Hill neighborhood of Seattle, WA had started using an e-bike for delivery. Lau interviewed the manager, Emily, who stated that: "my staff feels more confident on this bike . . . they can concentrate on the traffic better—enabling them to be safe. It really handles our heavier deliveries with ease." She went on to state "We only have delivery at our Capitol Hill location. If we open our Ballard location up to delivery, we would certainly buy an electric bike for that purpose" (Lau 2013).

Facility design

Much of the American urban environment has been shaped and defined by automobile infrastructure. Automobiles require a large impervious surface. They must be stored on such a surface at every destination. The city and countryside have been entirely fragmented with broad corridors of pavement, including interstate systems, local streets, and every other type of roadway facility. Bikes, on the contrary, need far less space.

Clearly, the provision of appropriate infrastructure has an effect on ridership, safety, and the viability of the e-bike. In Amsterdam, a city with a 50 percent bike modeshare, there are 400 km (248 miles) of separated cycle tracks and cycle ways; and each year there are only around six fatal bike-related accidents (Parker 1999). The City of Portland, Bureau of Transportation (PBOT), is a leader in bike planning and design nationwide. As part of the Portland Bicycle Plan for 2030, the City produced a Bikeway Facility Design, Survey of Best Practices (Portland 2010). The bike facility designs discussed below come from that report and have been assessed herein, for their suitability for e-bikes (Portland 2010, Appendix D). PBOT's Best Practices include numerous bike-friendly design ideas from sharrows to cycle tracks. For e-assist bikes with the same performance and dimensions as conventional bikes, the need for new designs is low. But for the e-power bikes with greater speeds, there is a need for separated facilities that enable fast commuting and meandering strolls along the same route. The following design treatments, pulled from PBOT's Best Practices, have been considered for their suitability to e-bikes. The following design treatments have been selected from the Survey of Best Practices, and are among the designs that have differentiating implications for e-bikes.

Figure 6.4a E-trike hauling exhibit booth set-up for conference.

Figure 6.4b Cargo bike in China.

Bike passing lanes

Bike passing lanes allow two speeds of cyclists to coexist without conflict. These designs were originally conceived to simply help faster cyclists separate themselves from slower cyclists. This can help break up bike platoons and reduce frustration among faster cyclists. PBOT design guidance includes a minimum passing lane width of 5 feet adjacent to a 5-foot bike lane and a 4-inch wide skip stripe separating the lanes.

Cycle tracks

Cycle tracks provide added safety for all cyclists, physically separating them from motor vehicle traffic. The separation can be achieved with space or physical barriers such as raised concrete, planted islands, parked cars, or a painted barrier. While these facilities provide safe buffers for all cyclists, they actually inhibit movements between the auto lanes and the bike lanes. This inhibits passing, and thereby forces slow and fast bike commuting to co-mingle. Cycle tracks are present in a few European nations, as well as the cities of Cambridge, MA; Seattle, WA; St. Petersburg, FL; and New York, NY in the U.S. Cycle-track-like facilities are quite prevalent in China, though they are used by many different types of vehicles and for many different purposes. Ingeniously, some cycle tracks in China have heavy, but impermanent barriers separating them from the vehicle lanes (Figure 6.6). PBOT's

Figure 6.5 Hawthorne Bridge passing lane in Portland.
Source: Madi Carlson.

Figure 6.6 Cycle-track-like amenities in Zhouzhuang, China.

design guidance suggests a range of widths for cycle tracks. With wider tracks, passing does become an option.

Bike boxes

Bike boxes prioritize bicycle movements at intersections. The cyclist is visible to motorists by being in front of them, which is also a space relatively free of vehicle exhaust fumes (PBOT 2010). The bike boxes are largely unnecessary for e-bike commuting, and oddly suited to them at the same time. The bike box allows the bike a chance to overcome inertia and gain speed before the platoon of cars fills the street ahead. E-bikes very easily overcome inertia and are capable of much quicker starts than is common on conventional bikes. However, one criticism of bike boxes is that the boxes fill with bikes, in front of the automobiles, but then the bikes are immediately overtaken by the quicker autos, and forced to line back up in the bike lanes. In an urban environment with posted speeds of 20 or 25 mph, the e-bike will not be overtaken by the cars, and can precede the auto platoon to the next stop. This limits delay experienced by the automobile drivers as well as conflict points as bikes assemble in the boxes only to line back up again, shortly after the traffic signal turns green.

Separating users

Ideally, fast and slow users of a facility should be separated. Modes can be separated, which is most easily communicated and enforced, or speeds can be separated. If there is sufficient space, more modes are accommodated by three or four lanes, allowing a greater range of speeds. For separated lanes, PBOT's survey shared the following recommendations from the Dutch Centre for Research and Contract Standardization in Civil and Traffic Engineering. First, ideally, pedestrians and bicyclists should be separated. Second, bicycle and pedestrian traffic can be combined if there are fewer than 200 pedestrians per hour per meter of profile width (< 200 peds/hr/ meter). Third, full combination is possible at 100 pedestrians per hour per meter of profile width. Fourth, visual separation (simple marking) is effective up to rates of 160 pedestrians per hour per meter of profile width. Fifth, "soft separations" created by separating bicycle travel paths using different paving materials are recommended up to 160–200 pedestrians per hour per meter of profile width (PBOT 2010, 18).

The City of Portland suggests a minimum path of 12 feet and a standard width of 16 feet. The PBOT preferred width for a shared-use pathway is a 16-foot bike lane, centered in the right of way, with pedestrian paths on each side with 6-foot minimum width, and an 8-foot standard width (PBOT 2010). See Figure 6.7.

Figure 6.7 Separated uses in New Orleans' Audubon Park.

Conclusion

The electric bicycle (e-bike) constitutes a nearly new mode of transportation, a mode between other modes. Planning and facility design that recognizes the growth in e-bikes is called "middle modalism." Middle modalism demands the development of new designs for transportation facilities of various types and speeds. Failure to properly plan for the proliferation of e-bikes will result in conflicts among users, unsafe conditions, and may result in the banning of e-bikes, despite how well e-bikes are aligned with widely adopted goals for mobility, active transportation, and climate change. City leaders, transportation advocates, bike planners, and engineers need to acknowledge and thoughtfully consider middle modalism in transportation planning and design.

Acknowledgments

The authors would like to thank the following individuals for their support and assistance. Nick Tahran, of Otak Inc. in Portland, provided a much-needed edit and other assistance with the production of this chapter. Alan Snook, of DKS and Associates in Portland, is partly responsible for the idea of using the SW Corridor time-shed modeling to portray the range of e-bikes. Alan Bassok introduced the authors and helped coordinate their work together.

References

Castellan-AG. 2012. *Market Study on E-mobility*. Kreuztal, Germany: Castellan-AG.

Cherry, Christopher R., Jonathan X. Weinert, and Yang Xinmiao. 2009. "Comparative Environmental Impacts of Electric Bikes in China." *Transportation Research Part D* 14(5): 281–290.

Dill, Jennifer and Geoffrey Rose. 2012. "E-bikes and Transportation Policy: Insights from Early Adopters." *Transportation Research Record: Journal of the Transportation Research Board* 2314: 1–6.

EV World. 2013. *EVELO 30-Day Challenge: Live Without Your Car, Ride E-Bike Instead*. 31 March. http://evworld.com/news.cfm?newsid=29919.

Fairley, Peter. 2005. "China's Cyclists Take Charge: Electric Bicycles Are Selling By The Millions Despite Efforts To Ban Them." *IEEE Spectrum*, 1 June. http://spectrum.ieee.org/transportation/advanced-cars/chinas-cyclists-take-charge.

Galbraith, Kate. 2012. "Batteries Add Power and Market Appeal to Bicycles." *New York Times*, June 28. www.nytimes.com/2012/06/28/business/energy-environment/batteries-add-power-and-market-appeal-to-bicycles.html.

Geller, Roger. 2006. *Four Types of Cyclists*. Portland Bureau of Transportation, Portland, OR. www.portlandoregon.gov/transportation/article/158497.

Goes, Hans. 2009. "The Silent Revolution." *Eurobike Show Dailies*, September.

Hendriksen, Ingrid, Luuk Engbers, Jeroen Schrijver, Rene van Gijlswijk, Jesse Weltevreden, and Jaap Wilting. 2008. "Rapport Elektrisch Fietsen—Marktonderzoek en Verkenning Toekomstmogelijkheden." *Electric Cycling: Market Research and Exploration of Prospects*. Brussels: Kwaliteit van Leven.

Lau, Cathy. 2013. "Seattle Cafe Delivers Tasty Grub with Electric Bike." *Electric Bike Report*, 17 May. http://electricbikereport.com/seattle-cafe-delivers-electric-bike/

Meyerson, Hilary. 2013. "E-Bikes: Hill-Flattening Roadsters Coming Into Their Own." *Outdoors NW*. http://outdoorsnw.com/2013/e-bikes-hill-flattening-roadsters-coming-into-their-own/.

National Household Travel Survey. 2010. *Highlights of the 2009 National Household Travel Survey*. Washington, D.C.: U.S. Department of Transportation.

Navigant Research. 2012. "Electric Bicycles: Global Market Opportunities, Barriers, Technology Issues, and Demand Forecasts for E-Bicycles, Pedal-Assist Bicycles, and E-Bicycle Batteries and Motors." *Navigant Research*. www.navigantresearch.com/research/electric-bicycles.

Parker, Alan A. 1999. "Power Assisted Bicycles Flatten Cities." *Australian Cyclist*, February/March, 61–63.

Portland Bureau of Transportation (PBOT). 2010. *Bikeway Facility Design: Survey of Best Practices*, Appendix D. Portland, OR: PBOT.

Weinert, Jonathan Xavier. 2007. "The Rise of Electric Two-Wheelers in China: Factors for their Success and Implications for the Future." *Institute of Transportation Studies*, University of California, Davis. Research Report UCD-ITS-RR-07-27.

7 Why we should stop talking about speed limits and start talking about speed

Arthur Slabosky

There has been a sea change in the Transportation Engineering world since people introduced traffic calming and other ideas for reducing the omnipresence of automobiles in our culture. The Institute of Transportation Engineers (ITE) now welcomes luminaries of the walkability movement onto its boards. The latest manuals now contain designs for bike lanes and for effective pedestrian crossings and road diets. It is now acceptable engineering practice to limit lane widths to the narrowest feasible, and to foster speeds that may not deliver the highest number of motor vehicles through a particular cordon line during a given length of time, to facilitate such things as bicycle and pedestrian (bike-ped) friendly design.

Yes, much has changed in the transportation design world. Bike-ped advocates get along with engineers. In fact, many of the bike-ped advocates *are* engineers. Traffic engineering for walkable communities is now serious business. However, in contrast to the many elements of practice that have changed for traffic engineers, one element has held fast, to the befuddlement of many in the sustainable transportation movement. This thing, more than even issues about separate bike lanes, raises more ire than all other issues combined: how we set *speed limits*. There are several ways to express why the process for setting speed limits does not change. Fundamentally, it is because speed limits are not a cause; they are an effect.

The basis for setting speed limits in the United States is a measure that is often referred to as the "85th-percentile rule." This means that if 85 percent of drivers are driving up to and including x miles per hour, we set x as the speed limit, usually rounded to the nearest, or to the next highest increment of 5 miles per hour. Critics of this method almost invariably want the speed lower than x. They believe that it is wrong to allow drivers to drive legally up to x, when all of us, presumably, should know that speeds, on said road corridor, ought to be lower on behalf of a better life for all users. The engineer's counterclaim to this argument is that there is no use in posting a restriction that you cannot enforce. Nor do critics frequently present a useful alternative to the 85th-percentile rule.

In an attempt to disengage the heretofore inevitable arguments on this issue, in this chapter I share conversations that I have had with people over

17 years of participating in bike-ped advocacy as well as advocacy for scientifically set speed limits. If we can find the reasons for intransigence about the matter, we may be able to find the arguments needed to bring critics into the 85-percentile fold. The first item that can sway an intransigent is an issue of fact: It is a fact that simply posting a sign does not change the travel speeds on a road. In short: speed limit is one thing; speed is something else.

The very people who are upset that we do not change speed limits are the same people who complain "everybody drives over the speed limit." This looks like a clear argument for some of us that speed limits are limited in their power. But the most telling aspect of the assertions made by critics are the comments that they themselves make: For instance: "it's a shame that we don't feel comfortable walking across our city streets because they are built for speed" and "NOBODY follows the speed limit." Now it all fits.

Once you say that nobody follows the speed limit, you must face the fact that establishing a speed limit is not going to do much to effect the driving behavior that you want to create. And once you acknowledge that, you have to admit that an unwillingness to establish a low speed limit is not a refusal to endorse traffic safety but an acceptance of reality—then, when we all agree that we forget about speed limit as a strategy, we can pursue engineering methods to reduce travel speeds. We have books full of such methods. When we are finished shaping the desired speed of traffic, we can set the speed limit at the 85th-percentile speed and be happy with it.

This is not to say that there are not legitimate disputes about speeds. There are relevant arguments about the roles of various streets and who belongs on them, and at what speed. This is where the arguments should be: about the speed itself but not whether the signs should indicate the speed that you have to stay below. One blogger asked me "if speed limits are useless, what are all these '20-is-enough' zones about?" The answer is that the promoters of such zones should be seriously skeptical about the probability that they can effect 20 miles per hour (mph) on the multitude of residential streets in U.S. neighborhoods. With the relatively low population densities in the majority of our neighborhoods, there is not enough tax base per block to justify cities investing in engineering design work to develop residential streets for slow speeds. These same low densities also force us to drive; generating too many car trips to allow us to turn our residential blocks into slow-speed zones.

While challenges related to low densities exist, it is possible to create very low travel speeds on city streets through engineering solutions. San Miguel de Allende, Guanajuato, Mexico, has such a speed and a speed limit—of 12.4 mph (20 kilometers per hour) as well. Streets are narrow with residential walls coming right up to sidewalks. Most of the population walks, and there are speed humps at every corner. Streets have a rough surface because they are made of stone. In this environment, the bike-like speed seems completely normal even to a vehicle occupant. No one has to question the speed limit there; that's the speed you drive because it's as fast as you *can* drive.

What are speed limits for then? Stakeholders of the transportation environment give different explanations. First, the raw speed of vehicles is not a major factor in crashes. Rather, it's differences in speed that matter. This is a favorite. Although it makes sense, studies have not clearly demonstrated this to be true. This explanation is also not sensitive to non-motorized travel modes because we know that every increase in speed bears a higher fatality risk to bodies that are not surrounded by steel. Second, speed limits may "set off" deviant drivers, i.e. they may speed in response. This is true, but it does not assuage people who think that everyone in a driving environment is going too fast. Again, every driver may be going too fast for the comfort of non-motorized users in the subject environment. Third, speed limits inform the unfamiliar driver, pedestrian, or cyclist what the normal vehicle speed is. This is important. It is no use "sending a message" that you would prefer a lower speed on the road if you cannot enforce it.

In the end, it is some of all of the above. The purpose of a speed limit is to present an enforceable level of vehicular speed. When such a distinction occurs with only 15 percent of drivers in non-compliance with the law, the fastest drivers for the most part can adjust their speed downward because they find the limit to be within reason. As well, the slowest drivers adjust their speed upward to meet the speed limit, bringing them closer together to the main cohort of traffic. This is why the 85th-percentile rule works.

Do speed limits control speed?

Speed limit is *not* a design element in and of itself; it is a result of design elements. James Baxter, founder of the National Motorists' Association states that "Attempting to dictate travel speeds through the use of speed limits, has not and will not work, and will, in all likelihood, diminish traffic safety" (Baxter n.d.). The Institute of Transportation Engineers (ITE) reports that "it is important to note that setting speed limits lower than 85th percentile speed does not encourage compliance with the posted speed limit" (ITE 2012). The Federal Highway Administration's (FHWA) Pedestrian Action Plan notes: "If 85% of motorists are doing the right thing, then enforcement can effectively manage the other 15%" and "If 85% of motorists are doing the wrong thing, then enforcement can do little to change behavior, especially if the built environment contributes to the illegal behavior" (FHWA 2007). And, finally, from Eric Dumbaugh, a practitioner and theorist in the field of walkable streets: "There is growing evidence that suggests that even conscientious drivers have difficulty obeying safety information displayed through various sign applications" (Dumbaugh 2005, 36–37). Dumbaugh goes onto summarize remarks by other authors in the field, writing: "in the absence of aggressive law enforcement, drivers will increase their operating speed to the 'safe' speed they infer from a roadway's design" (130).

These sources represent the full range of the political spectrum in transportation from the "cars rule" people to the "complete streets" people. The NMA is very unfriendly (or indifferent) to non-motorized travel while Eric Dumbaugh is very friendly. The other entities fall somewhere in the middle. Nevertheless, above all, these sources agree that drivers' travel speeds are a result of the travel environment and not a reaction to text on signs. In contrast, some advocates of complete streets do not subscribe to the above statements and take umbrage with the above ideas and to the professionals that promulgate them. For example, "One for the dustbin: the 85th Percentile Rule in Traffic Engineering" (Schmitt 2012); and "Have any white papers or other documents . . . been prepared to show the idiocy of using the 85th percentile to set speed limits" (Association of Pedestrian and Bike Professionals 2008); in reference to the 85th percentile rule, "The old 85th percentile trap . . . a well-greased, dark, sleezy [sic] slipper-slimey critter inhabiting the oldest of the halls and catacombs of reverse gear thinking" (Burden 2004). Such advocates are properly concerned about having reasonable speeds on streets that are friendly to all kinds of users. But they present a debate that is, perhaps, unnecessarily partisan and frustrate transportation professionals who are needed to implement the results that they desire.

It is not unusual in many realms of political life for people to feel as if you are "giving in" if you face reality. An apt comparison to the speed limit argument is in approaches to the use of prohibited drugs. Hard-core advocates of prohibitionist policies will not accept that laws that prohibit consumption of certain kinds of substances do not prevent people from engaging in such consumption.

While we do not have a definite method to reduce drug consumption, we do have specific methods to reduce motor vehicle travel speeds. Recognizing that posting of signs does not work is the first step to proceeding to methods that do work. Unfortunately, public officials have an incentive to promote unhelpful sign-message answers to high travel speeds; when every motorist is breaking the law because they are all exceeding the speed limit, every motorist is subject to apprehension and ticketing. This brings in easy revenues to cities, without any need to distinguish very fast deviant drivers from the rest. In other words, actual safety takes a back seat to imagined safety as well as revenue production. Figure 7.1 depicts a typical example of a political change, one that occurred in the community of Okemos, Michigan. After a high-school-aged driver was killed in a car accident as he tried to enter Jolly Road, there was a public outcry that led to a reduction of the speed limit to 45 miles per hour from 55 miles per hour.

The speed profile from before and after is virtually unchanged. The highest speed, the 50th percentile and the 85th percentile are virtually unchanged.

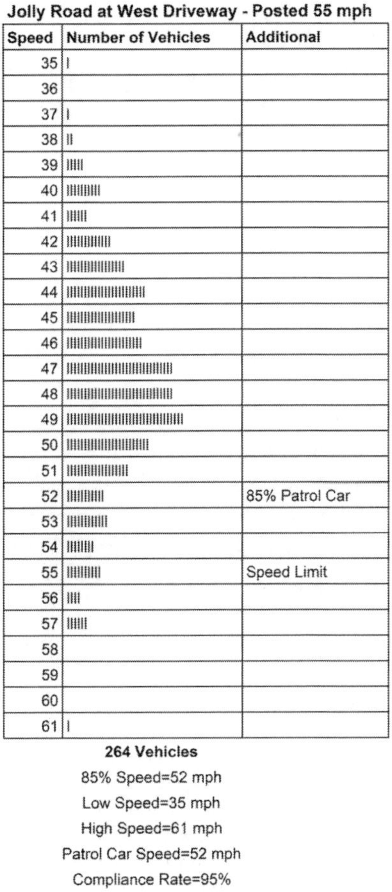

Jolly Road at West Driveway - Posted 55 mph

Speed	Number of Vehicles	Additional
35	I	
36		
37	I	
38	II	
39	IIII	
40	IIIIIIIII	
41	IIIII	
42	IIIIIIIIII	
43	IIIIIIIIIIII	
44	IIIIIIIIIIIIIIIIII	
45	IIIIIIIIIIIIIIII	
46	IIIIIIIIIIIIIIII	
47	IIIIIIIIIIIIIIIIIIIIIII	
48	IIIIIIIIIIIIIIIIIIIIII	
49	IIIIIIIIIIIIIIIIIIIIIIIIII	
50	IIIIIIIIIIIIIIIIIII	
51	IIIIIIIIIIIIIII	
52	IIIIIIIIII	85% Patrol Car
53	IIIIIIIIII	
54	IIIIIII	
55	IIIIIIII	Speed Limit
56	IIII	
57	IIIII	
58		
59		
60		
61	I	

264 Vehicles
85% Speed=52 mph
Low Speed=35 mph
High Speed=61 mph
Patrol Car Speed=52 mph
Compliance Rate=95%

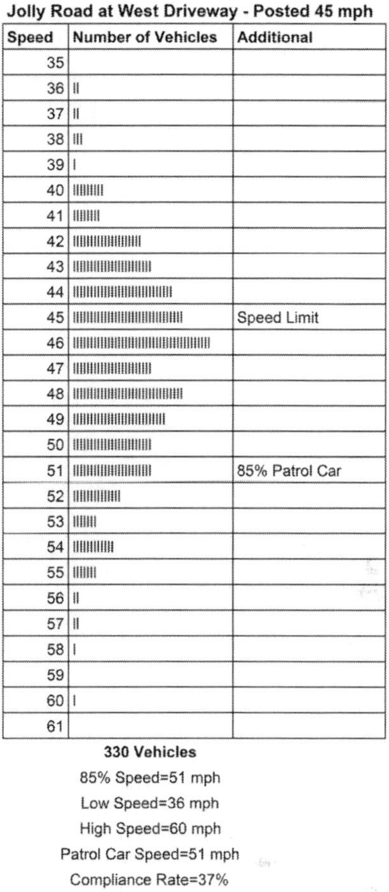

Jolly Road at West Driveway - Posted 45 mph

Speed	Number of Vehicles	Additional
35		
36	II	
37	II	
38	III	
39	I	
40	IIIIIIII	
41	IIIIIII	
42	IIIIIIIIIIIIIII	
43	IIIIIIIIIIIIIIIII	
44	IIIIIIIIIIIIIIIIIIIIII	
45	IIIIIIIIIIIIIIIIIIIIIIIIII	Speed Limit
46	IIIIIIIIIIIIIIIIIIIIIIIIIIIIII	
47	IIIIIIIIIIIIIIIIIII	
48	IIIIIIIIIIIIIIIIIIIIIII	
49	IIIIIIIIIIIIIIIIIIIII	
50	IIIIIIIIIIIIIIIIII	
51	IIIIIIIIIIIIIIIIII	85% Patrol Car
52	IIIIIIIIIII	
53	IIIIII	
54	IIIIIIIIII	
55	IIIIII	
56	II	
57	II	
58	I	
59		
60	I	
61		

330 Vehicles
85% Speed=51 mph
Low Speed=36 mph
High Speed=60 mph
Patrol Car Speed=51 mph
Compliance Rate=37%

Figure 7.1 Before-and-after charts from the state police speed study.
Note: The right chart is denser than the left because there were more cars in the later study.

The only difference is legal. Where 15 percent of the drivers used to be violators, now 63 percent are violators. Instead of isolating the deviant drivers, it is now "normal" to be in non-compliance. This is a real life illustration of the earlier comments about signs not controlling speed. And now police can apprehend and fine their choice of two-third of the drivers on Jolly Road, whether they are posing a safety hazard or not. For the public, the problem may appear to be solved. They asked for a lower speed *limit* and they got it. They just asked for the wrong thing. And the local jurisdiction is glad to collect the fines.

It is important to note that the fact that speeds remain the same on the subject street does not mean that there is no effect because of the speed limit change. A frequent effect is a change in attitudes. Namely, when the community decides that the speed limit reduction is a solution to the perceived problem, they often stop there without looking for measures that actually may have long-lasting benefits. In the above example, there was no effort to calm traffic at the site.

East Lansing I-69 case study

The following case study wraps up this chapter and illustrates how *raising* the speed limit once again does not change speeds, but can have positive side benefits. The case begins in the 1970s, sadly, with another teenager's death. After a fourteen-year old was hit on his bike and killed by a car on Temporary I-69 (now Business I-69) in East Lansing, Michigan, the state lowered the speed limit on that road from 45 miles per hour to 35 miles per hour. This speed limit remained posted until 2007. During the roughly three decades that the 35 mph was posted, the bulk of drivers were observed driving in the 40 mph range. Extant data do not exist to show how or whether the crash record changed. In 1989 the I-69 freeway was completed past East Lansing which diverted large trucks off the Business route, further diminishing any possible basis of comparison before and after the speed limit change.

In 2000, I was on the City of East Lansing Transportation Commission and suggested that we ask for a traffic study to see whether speeds actually reflected the speed limit along that stretch of road. One of my fellow commissioners, a former mayor, did not want the study. He said that he didn't "want to go there." The majority of the commission, however, voted for the study. When the state police concluded that the 85th-percentile speed was in the mid-40s, the speed limit was increased to 45 mph. This lasted for four months, while the city mounted political pressure against the state. The city even claimed that the request for the speed study was invalid because the transportation commission was an advisory body and did not have the authority to make official requests. City council may have been correct about that; they never went to court on that issue. But consider: A city was asking to discard on-the-ground facts that they would just rather not know! To the city council it was okay to present a false picture of the situation. In fact, under the 35 mph speed limit, only 3 percent of drivers were at that speed or lower. This is the information that the city council did not want the general public to know. The story of what happened next will show why deliberate ignorance is not the best answer for safety.

In 2006, the Michigan State Legislature established a law mandating that speed limits be set according to engineering studies. This meant that according to that law, the Business I-69 speed limit should be set at 45 mph. The State posted the 45 mph speed limit in January 2007 over severe protests from the East Lansing City Council, and a lawsuit resulted, which the state won. What followed was: First, in 2008, crime in the city reached its lowest point in 28 years. Second, in 2010, parts of the corridor received a sidewalk on both sides of the highway where one had never been before. And, third, crashes in the city went down by 20 percent (as calculated by the author via a public crash calculation utility provided by the Michigan State Police).

The sidewalk was a political action that was demanded when people were forced to face the truth about the traffic situation; when the *perceived* speed was 35 mph, the public accepted that the situation was under control. When the perceived speed went to 45 mph (even though actual speed did not change) the aggrieved city persuaded the state to provide a million dollars in funds to build a sidewalk that the city had never felt necessary before. If the public had responded to the bicyclist's death in the 1970s with the idea that a sidewalk would help, there could have been one 30 years earlier.

The reduction in traffic crashes throughout the city is arguably the result of police concentrating on activities besides speeding on Business I-69. We must keep in mind that there are many more driver violations and causes for bike-ped discomfort and injuries than speeding. Among them are right-on-green crashes and general denial of right-of-way to pedestrians. These violations are hard to get enforced when there is a steady supply of speed violators available with low speed limits. Astute readers may notice that the recession broke loose in 2008, with an accompanying reduction in traffic. This reduction, however, was about 4 percent and could not alone justify a reduction of 20 percent in crashes. Another figure reported in the local newspaper was that speed tickets issued went down by 378 from 2007 to 2008. This is telling evidence that speeding enforcement did in fact decline. There were no layoffs in the city police force. The police must have gone somewhere else, probably where the 20 percent fewer crashes occurred.

More about speed limit as "the Truth"

Speed limit may only be a concept, but it is an item of information that is posted on every major road segment in our country. When our citizens see speed limit signs, they want to be able to use that information as a guide to what speed drivers are actually travelling. We do not want to have to carry out a speed study every time we consider something on a road; it is impossible to do a speed study everywhere. Consider data about crashes versus speeds. Speed limit is frequently used as a proxy for speed in research focusing on traffic crashes and their causes. Because speed limits are often set low relative to actual average speeds, we do not always know what the actual travel speeds are when conducting investigations. Hence we are no doubt lumping together many road segments that officially have the same "speeds" that may have very different actual speeds, and vice versa. If we had nationwide consistency of speed limits that reflected speeds, we would have a reliable base to start with to determine crash causation in terms of speed.

We also have rules about placing certain elements together that belong with certain kinds of speed ranges. We have minimum requirements for overpasses and maximums for parklets, bike lanes, crosswalks, and other bike-ped features. Proper yellow signal interval timing depends on knowing the speed of a facility. There are times that we have to rely on speed limit as a surrogate for speed. Also, the public has a sense of an acceptable speed for its own pedestrian crossing activity or bicycling in the road. It is proper and informative for the number on a sign to present accurate information with respect to the actual travel speeds. This is what our users want.

Conclusion: the role of speed limits

Speed limit signs do not significantly alter the speed of drivers. Lowering a number does not inherently reduce driver speed nor does raising the number inherently raise driver speed. Setting a speed limit recognizes the reality of the current road configuration and environment. It is not a declaration that such speed is the best for that facility in perpetuity. Changing driver speeds requires tangible engineering changes to the design of the facility. When such changes are implemented to bring about a desired speed pattern, the desired speed limit follows. Posting of speed limits below the 85th percentile offers the following risks: First, enforcement efforts concentrate on a large cohort of reasonable drivers, to the detriment of enforcement of more serious driver misbehaviors. And, second, citizens who interact with the facility are dangerously presented with a false notion of the speed of car traffic.

Any of a whole set of facilities that use a speed rule as a guide to placement may be improperly placed or improperly rejected, or placed and designed improperly because the speed limit information does not reflect actual speed. Determining and posting speed limits according to the 85th percentile of vehicle speed achieves the following: First, it reveals real deviant drivers who warrant public enforcement. Second, it helps to present a realistic picture of the traffic situation to all users. Third, this provides a shared starting point to all interested parties for further discussion of how the traffic environment might be altered to the benefit of the community. A properly set scientific speed limit is the best way to start on the path to re-engineering a road corridor to the speed that you want to put into effect. When you effect that speed, you will be glad to have an 85th-percentile rule to provide a clear criterion as to the posted speed limit. Now both the speed and the speed limit are what you desired.

References

Association of Pedestrian and Bike Professionals. *Listserve post*, 15 June 2008.

Baxter, James. (no date). "The Truth About Speed Limits." *National Motorists Association*. www.motorists.org/issues/speed-limits/truth/.

Burden, Dan. 2004. Personal communication. 2 March.

Dumbaugh, Eric. 2005. Safe Streets, Livable Streets: A Positive Approach to Urban Road Design. PhD diss., Georgia Tech.

Federal Highway Administration (FHWA). 2007. *Developing a Pedestrian Safety Action Plan*. Instructional Course. Flint: FHWA.

Institute of Transportation Engineers (ITE). 2012. *Methods and Practices for Setting Speed Limits: An Informational Report.* Washington, D.C.: Federal Highway Administration. http://safety.fhwa.dot.gov/speedmgt/ref_mats/fhwasa12004/.

Madruga, Pedro. 2012. "The 85th Percentile Folly." *Copenhagenize.com*, November 16. www.copenhagenize.com/2012/11/the-85th-percentile-folly.html.

Schmitt, Angie. 2012. "One for the Dustbin: The 85th Percentile Rule in Traffic Engineering." *Streetsblog USA*, 16 November. http://usa.streetsblog.org/2012/11/16/one-for-the-dustbin-the-85th-percentile-rule-in-traffic-engineering-2/.

8 A framework to analyze the economic feasibility of cycling facilities

Mingxin Li and Ardeshir Faghri

Introduction

It has been shown that participation in regular physical activity, such as active travel through walking and cycling, reduces the risk of coronary heart disease, stroke and hypertension, type II diabetes, obesity, as well as some types of cancer and depression (World Health Organization 2010). A significant and growing body of evidence links insufficient physical activity to many medical problems (Olabarria et al. 2012; Rutter et al. 2013) and modest increases in physical activity have the potential to produce substantial health benefits (Haskell et al. 2007).

Non-motorized mobility options, such as bicycling and walking, described as "the forgotten modes" of transportation, have been overlooked by many federal, state, and local agencies for years (Wilkinson 1997). Since the early 1990s, a growing literature has been devoted to non-motorized transportation (Taylor 2009; Turner, Hottenstein, and Shunk 1997; U.S. Department of Transportation 1999). Lusk et al. analyzed and compared U.S. bicycle facility guidelines published between 1972 and 1999 (2013). They concluded that, in the United States, the rate of vehicle–bicycle crashes on cycle tracks was lower than published rates for bicyclists on roads.

As an essential pre-requisite to make the best use of limited transportation funds, economic assessment—primarily cost–benefit analysis (CBA)—is a fundamental part of the specialized approach of a variety of professionals including transportation planners, policy advisors, and environmental project managers. Cost–benefit analysis is a family of techniques designed for appraising and determining the feasibility of investment projects by systematically quantifying their costs and benefits. Therefore, cost–benefit analysis is not simply a method of determining the least-cost alternative, but an approach to identify the most cost-effective alternative based on considerations of cost, benefit, and confidence factor. CBA has become a powerful tool used in planning and policy efforts at all levels of transportation planning.

Economic analyses of transportation investments are published with increasing frequency in the transportation research literature. The majority of the research interests in the field of CBAs have been concentrated on assessing the impacts of road investment projects in the USA and many other countries. Since transportation planners have been faced with considerable challenges in trying to estimate the benefits of bicycle facilities, a favored approach to estimate the cost of various bicycle facilities involves cost–benefit analysis of cycling policy options incorporating estimation of both direct benefits to the users of the facilities and indirect benefits to the community (Krizek 2007).

The *Wisconsin Bicycle Facility Design Handbook* (2004), updated in 2006 and 2009, contains detailed guidance for the development of bikeways. Korve and Niemeier (2002) developed a cost–benefit analysis framework to analyze the impacts of improved bicycle phasing at an existing signalized intersection. They then conducted a sensitivity analysis to see how the cost–benefit relationship would vary according to the discount rate used. Salensminde (2004) presented a CBA approach that took into account the benefit of reduced insecurity and the health benefits of the improved fitness the use of non-motorized transport provides. The results show that the benefits of investments in cycle networks are estimated to be at least 4–5 times the costs. The guidelines for analysis of investments in bicycle facilities provide a detailed specification of the categories of cost and benefits to be included in analyses for cycling infrastructure projects. While these ingredients are provided, it is still unclear how to evaluate health-related economic aspects for existing and new cyclists.

Krizek et al. (2006) examined a range of efforts to increase bicycling in the United States and provided methodologies to estimate the preliminary cost of different types of bicycle facilities and developed criteria for identifying benefits associated with bicycle-related projects. They also estimated the demand for a cycle-way network and provided an overview of literature on the demand element that is an important part of input data for the CBAs. Krizek (2007) further reviewed and interpreted the literature that evaluates the economic benefits of bicycle facilities. Krizek's study extends the earlier work to discuss the appropriate use of their methods and quality of the economic analyses. Pucher and Buehler (2006) examined a range of possible causes of the bike share of urban trips in Canada compared to the United States. They use multiple regression analysis to explore the relative importance of each impact factor. They concluded that higher densities and mixing of land uses probably encourage more cycling.

Motivated by a desire to reduce the prevalence of global obesity and environmental problems, international organizations have been calling for multidisciplinary approaches to increase physical activity and reduce reliance on cars and their environmental consequences. In 2006, the World Health Organization (WHO) regional office for Europe undertook a project on economic valuation of health effects from cycling. This office (Cavill et al. 2007, 2008; Rutter 2007) developed the Health Economic Assessment Tool for cycling (HEAT for cycling) to estimate the maximum and the mean annual benefit and values in terms of reduced mortality as a result of cycling.

Gotschi (2011) presented a conceptual framework of the cost–benefit analysis to evaluate health related economic aspects of past and planned development of bicycling in Portland, Oregon. This was one of a small number of studies to include both monetized health benefits and health care cost savings in its calculations. The benefits of fuel savings are not included in this framework. In addition, this analysis does not take into account the costs of bicycle ownership and maintenance. More recently, Meschik (2012) split the total costs into internal costs primarily paid by transportation users and external costs from transportation, which are not paid by transportation users directly but by the general public, such as pollution costs, accident costs, and health expenditures. Accident costs were not calculated in this study.

However, many significant questions remain to be addressed, particularly regarding costs that should be monitored and recorded as well as quantitative and qualitative benefits that can be attained through investments in policies and initiatives, in particular those related to the implementation of sustainable transport policies, such as cycling. For example, one of the critical challenges concerning the economic appraisal of health effects related to cycling is the relationship between observed cycling and total physical activity—for example, increased cycling might increase a person's overall physical activity. Also, many consider spending public monies on cycling facilities a luxury due to lack of identified and quantified benefits in the current literatures. How to evaluate the benefit of new bicycle facilities and justify these investments, especially as compared to other modes? And how long does it take for an investment to pay for itself?

Thus, this chapter seeks 1) to establish a standard analytical approach for analysis and documentation, 2) to analyze cycling investment from an economic framework with cost–benefit setting, and 3) to provide planners, transportation specialists, and health advocates with detailed economic analysis worksheets to determine the full costs and benefits of promoting bicycling in urban and suburban areas.

The remainder of the chapter is organized as follows. We first discuss the proposed CBA framework by introducing the main inputs and outputs for estimating the most important benefit components and expenses associated with bicycle facilities. The following section presents a case study in which a cost–benefit analysis is developed to determine whether the benefits of an added cycling facility can justify the expenditure of several million dollars. Finally, the chapter makes recommendations for research, policy, and other measures that can be taken to meet the goals of the study.

Cost–benefit analysis framework

Investment projects such as adding new cycling facilities to increase levels of walking and bicycling may affect a wide range of parameters relating to economics, mobility, environment, and health. In order to identify outputs, for example, the cost and monetary benefits for society, it is necessary to enter characteristics about the size and type of a proposed facility as inputs before conducting a complete and comprehensive CBA. This determination has to cover all relevant aspects, including those on population, mode share rate, demand, human health, and the environment. Since there are no conclusive studies to guide us on the likely proportion of social benefits, such as improved livability or increased economic activity, cost–benefit analyses, generally, as well as this chapter, specifically, do not take into account social benefits per se, although some of the techniques described below could be adapted to do so. Planners and other transportation specialists might input minimal specifications to receive a simplified approximate estimate or enter highly detailed information that will allow for a more complete and accurate analysis (Figure 8.1). Valuation techniques are discussed and attempts to include the cost of the cycling project and health effects in a CBA of development of infrastructures for cyclists are reviewed.

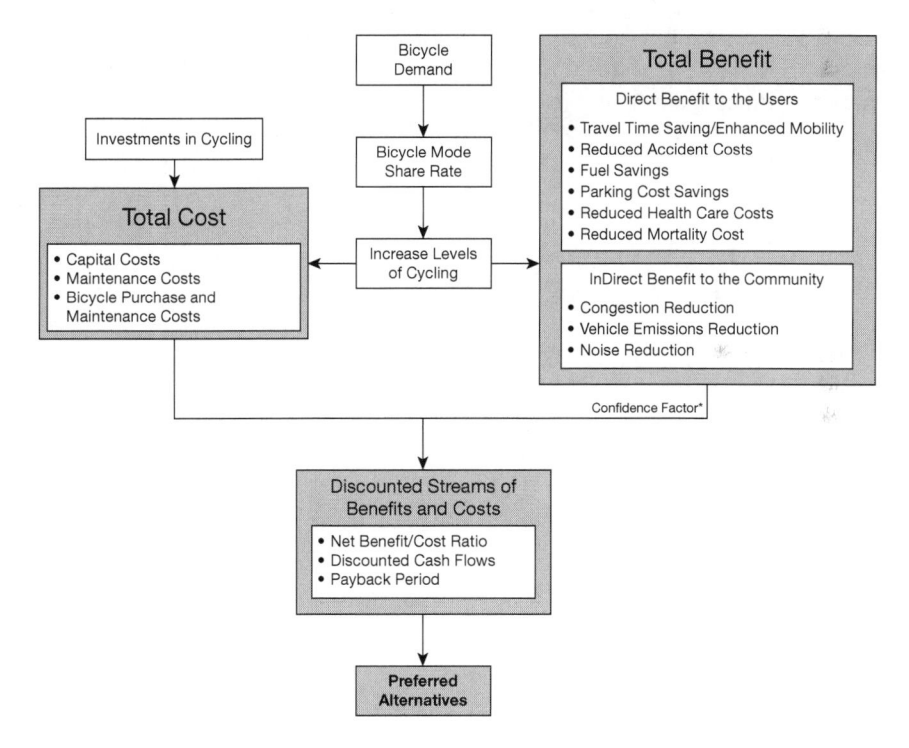

Figure 8.1 Conceptual framework of the cost–benefit analysis. For each benefit, a confidence factor is introduced as an indication of the degree of certainty in the estimates. For example, a 95 percent confidence factor indicates a high degree of certainty that the estimates are correct, while a confidence factor of 60 percent would indicate a lower degree of certainty.

Cost

To reasonably identify, apply, and project the costs for each alternative, the cost analysis provides a comprehensive estimate of the costs for cycling projects, which are typically broken into two major categories: capital costs and maintenance/operating costs including pavement, drainage, traffic controls, and landscape maintenance. The construction cost level depends on categories of bicycle facilities: on street, off-street, and equipment (Krizek et al. 2007). Cycling trail maintenance costs depend on numerous factors such as location, volume of use, types of use, pavement type, and level of voluntary participation.

Unit costs of bicycle facilities

Costs of items may vary depending on location and suppliers. Some unit costs are taken from "NYSDOT Quick Estimator Reference-Calculations–Upstate" 2012. As shown in Figure 8.2, the real costs of building highway and street facilities have been growing consistently over the last few decades. To avoid underestimating the needs of a critical project, it is necessary to make an inflation-adjusted estimate of the costs based on the inflation factors for material and supply inputs to Highway and Street Construction. All cost values were normalized to a base year of 2012. Inflation factors were developed based on Producer Price Index for Highway and Street Construction from the 1987–2009 data to convert unit costs from 2012 levels to the build year. Doing so will ensure each year's Producer Price Index (PPI) calculation (US Department of Labor 2013) is consistent with the inflationary pressures facing the recurring cost such as operating/maintenance cost for that year.

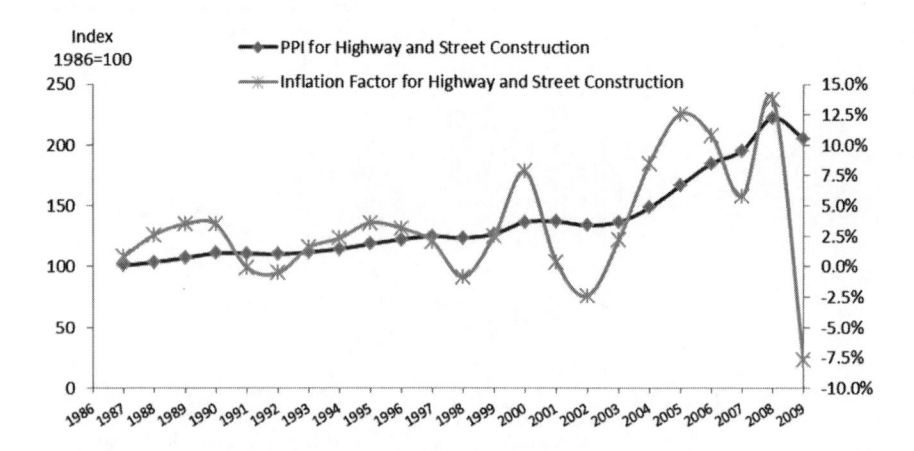

Figure 8.2 Producer price index for highway and street construction, 1987–2009.
Source: Data sourced from the U.S. Bureau of Labor Statistics.

Estimation of the costs in our framework involves several assumptions, including:

- Costs are based on standard facilities constructed in the United States and are represented in year 2012 dollars. They may change due to future economic conditions.
- Facility costs include construction and design.
- Discount rate = 5%, with a standard time horizon of 30 years.[1]

Available information on the unit costs of bicycle facilities was obtained from municipal transport documents, cost estimates of individual projects, and consulting reports. To normalize each cost element to a national level, Krizek et al. (2006) developed a construction cost index by state or region. The trail maintenance and operation manual published by the Rails-to-Trails Conservancy (2005) provides cost information on maintenance and operation. Through a survey of 100 trail providers, this study found that 60 percent of trails reporting costs were maintained primarily by a government agency with annual maintenance and operation costs over $2,000 per mile. The other 40 percent of trails were primarily maintained by a non-profit or volunteer organization with annual M and O costs under $700 per mile. The overall average cost of maintaining a trail among those surveyed was $1,500 per mile per year. Indiana Safe Routes to School (SRTS) Program provides "Estimated Unit Costs for Bicycle & Pedestrian Facilities" which were updated for 2012.

User costs for bicycle purchase and maintenance

According to a survey of North American bicycle commuters conducted by Moritz (1997), median bicycle purchase cost and annual maintenance expenses were $500 and $714 in 1996, respectively, compared to those of $400 and $344 based on the "How much do you spend on commuter biking per year?" survey on Bike Forums website (2005). All estimates are inflated to 2012 dollars using the consumer price index (CPI) for sporting goods including bicycles with an annual average increase of 0.4 percent between 1996 and 2012 (www.bls.gov). Assuming that the average bicycle commuter has been commuting by bike for 8.3 years (2005), the inflation-adjusted user costs for bicycle purchase and maintenance are $614 per person.

Costs not included in our model

This approach does not account for cost of cyclists due to increased ventilation rate when exposed to air pollution (Marshall, Brauer, and Frank 2009; Rojas-Rueda et al. 2011). Additionally, this approach does not account for opportunity costs, as there is no conclusive evidence to guide us on a likely proportion of opportunity costs. In this case, rather than risk over-estimating the cost with a conservative assumption, we will assume that there is no opportunity cost incurred in the new cycling facilities.

Benefit

Once the cost categories have been developed, the transportation specialists must identify the profiles of benefits that apply to each feasible alternative over the system's life. As described previously, the main categories of quantifiable benefits of the cycling project are: direct benefit to the cyclists themselves and indirect benefits to the community. Direct benefits include enhanced mobility, increase in physical activity, reduced health care, reduced mortality, fuel savings, and decreased crashes. Indirect benefits to the community include decreased congestion, reduced pollution, improved livability, and increased economic activity. The objective of the benefit valuation process is to present a reasoned prediction of the value of the benefit of new bicycle facilities and justify these investments. The benefit model provides a comprehensive estimate of benefits of the proposed bicycle facilities including direct benefits to the users of the facilities and indirect benefits to the community.

Reduced health care costs

The negative health effects of physical inactivity are paralleled by staggering economic consequences. Colditz (1999) attributed $257 per capita to physical inactivity in the United States in 1996. Pratt, Macera, and Wang (2000) found the annual cost directly attributable to inactivity in the U.S. is an estimated $330 per capita. Wang et al. (2004) estimated the per capita health care costs from cardiovascular disease attributable to inactivity to be $231 in 2001. Garrett et al. (2004) estimated total health plan expenditures attributable to physical inactivity were $83.6 million in a health plan population of 1.5 million members, or $56 per member, which is substantially lower than the former three estimations. Garrett et al. (2004) argue that their study included only medical costs associated with a subset of conditions, while the Pratt et al. (2000) considered all medical costs related to inactivity. To quantify reduced health costs associated with physical activity, we followed the approach presented by Gotschi (2011) for cycling projects. All estimates are inflated to 2012 dollars using the consumer price index (CPI) for medical care with an annual average increase of 4.55 percent between 1990 and 2012 (www.bls.gov). The estimated average inflation-adjusted per capita health care cost is $635.

Congestion reduction

In 2011, congestion caused urban Americans to travel 5.5 billion hours more than they would have without the congestion as well as to purchase an extra

2.9 billion gallons of fuel for a total congestion cost of $121 billion (Schrank, Eisele, and Lomax 2012). Empirical evidence indicates that the net benefits of a shift from driving to cycling are estimated to average 20¢ per urban peak mile (Litman 2004).

Reduced mortality cost

The WHO regional office for Europe developed the Health Economic Assessment Tool for cycling (HEAT for cycling) to estimate the economic savings resulting from reduced mortality due to regular physical activity from cycling. In this tool, the average saving from reduced mortality is one of the most robustly quantifiable health outcomes. Indeed, direct benefits on health are hard to measure. To this end, we used the Health Economic Assessment Tool for cycling (HEAT for cycling) model to estimate the health effects from variety of levels of cycling.

Reduced external costs of air pollution and noise

Automobile air pollution costs are estimated to be 10¢ per mile for urban-peak driving (Litman 2004). Noise reduction benefits from automobile travel shifted to cycling are estimated to average 3¢ per mile for urban-peak driving (Litman 2004).

Reduced parking costs

In addition to reductions in congestion costs, mortality costs, and external costs, the analyses also take into account reduced parking costs created by shifting from traveling by car to cycling. Parking costs are measured per trip or per day rather than by trip length. In this study, shifting from automobile to cycling is estimated to provide parking savings of $4.00 per day (Litman 2004).

Indirect benefits not quantified in our model

An in-depth analysis of the possible improvements in bicycle facility selection in terms of cost–benefit analysis should include all the aspects, if a user's preferred option is to be identified with accuracy. However, some indirect benefits of a project are not included in the analysis since they may be difficult to quantify. In many cases, certain conservative assumptions are necessary in order to improve the potential accuracy of the CBA. It should be noted that including the impact of improved community livability and increased economic activity in an economic appraisal would lead to a larger extent of uncertainty. Therefore, to avoid overestimating the benefits this paper does not account for the benefits of improved community livability and increased economic activity as a result of cycling.

Case study

To demonstrate the sensitivity of the model to a number of key explanatory variables, it is helpful to examine a case study, which is essential to understanding the applicability of both the methodology of economic appraisal and tool implementation. The results serve as an example for future applications of the tool to transportation agencies.

Study site and data collection

The Wilmington to New Castle Greenway (Figure 8.3) is the transformation of an abandoned railroad right-of-way into the Whittier Greenway Trail, a 6-mile recreational and commuter bikeway and pedestrian path. It will connect two major cities—Wilmington and New Castle, Delaware—almost entirely by off-road trail facilities. The Trail begins from SR 273 in the City of New Castle and travels through the Christina River north of Interstate 495, linking schools, homes, transit stops, parks, and shopping areas. The trail has very few at-grade road crossings, providing a lush greenbelt for walking and bicycling.

Figure 8.3 Wilmington to New Castle Greenway.
Source: Google Maps 2013.

Upon completion, this will result in a continuous trail from the City of New Castle to the intersection of Market Street and Martin Luther King Boulevard in downtown Wilmington. Cyclists on this highway will be able to travel between Wilmington and New Castle in about 20 minutes compared to 15 minutes of average travel time by car. The 6-mile long "bicycle highway" also provides increased safety for those who prefer to bike or walk to their destinations. Eventually, the trail will be part of the East Coast Greenway route through Delaware.

The cost estimates shown in Table 8.1 are based on the cost information of three trails (Struble Trail, York County Heritage Rail Trail, and Capital Area Greenbelt), chosen based on data availability and representativeness of typical trails for most human forms of transportation (e.g. cycling, walking, running, etc.) from a survey of 100 Rail-trails in USA (Rails-to-Trails 2005). The cost estimates were developed by calculating rough quantities based on the cost information of the three trails and applying unit costs. Costs were then translated into per mile or per category costs. All adjusted cost values were normalized to a base year of 2012. Inflation factors were developed based on the Producer Price Index for Highway and Street Construction from the 1987–2009 data to convert unit costs from 2012 levels to the build year.

Benefit and cost components of investments in cycling facilities

The results of the cost–benefit analyses presented in Table 8.2 are based on a discount rate of 5 percent and a 30-year lifetime for the projects. The estimated cost to implement this plan over 30 years is approximately $30 million (based on 2012 dollars). The plan cost includes approximately $19.5 million for off-road bicycle facilities, $1.2 million for bicycle facility maintenance, and $4.2 million for tax cost.[2] The level of investment that will be required in order to implement this project is relatively modest in comparison to other transportation facilities. The net benefit is the present value of the benefit components minus the total costs incurred in the project. The net benefit/cost ratio indicates marginal returns, namely, the benefit per cost unit. As shown in Table 8.2, the net benefit/cost ratio equals 0.67. Considering the economic and health results, which are depicted in Table 8.3, one can easily conclude that the new development of infrastructures for cyclists leads to appealing results for investments in cycling facilities. Hence, the investment in the Wilmington to New Castle Greenway seems to be beneficial to society indeed. To illustrate the financial results, the above case has been economically evaluated and compared using general profitability criteria such as discounted cash flow, net benefit/cost ratio (NBCR), and discounted payback period (DPBP). To evaluate the profitability of the different scenarios, we have set the lifetime of the cycling facility to 30 years.

Apart from the net benefit/cost ratio, we also calculate the discounted cash flow. In order to make better strategic choices, it is essential to analyze the

Table 8.1 Summary of capital and operational/maintenance cost. To quantify the benefits of physical activity, we followed the approach presented in the Health Economic Assessment Tool for Cycling Project (HEAT)

Cost components	Value in dollars
• Bridge over the Christina River • Bridge over Little Mill Creek • Bridge with open decking over wetlands	
Material costs (including three bridges)	13,000,000
• Design, 10%	1,300,000
• Mobilization, 10%	1,300,000
• Construction Management, 10%	1,300,000
• Contingency, 20%	2,600,000
Additional costs	6,500,000
Total capital costs	19,500,000
• Bridge inspection	44
• Bush hog	240
• Crosswalks	69
• Culverts	46
• Designated projects	2,007
• Fence repair	596
• Flower bed planting	79
• Gates	61
• Grade ditches	73
• Invasive pruning	160
• Invasive spraying	205
• Leaf removal	48
• Mowing	946
• Pruning	350
• Signage	172
• Storm damage	160
• Trail surface (asphalt)	78
• Trash	474
• Leaf removal	125
• Trimming	489
• Vandalism	206
• Weeding	130
• Maintenance of toilets	590
• Maintenance of informational kiosks	71
• Maintenance of pavement markings	67
• Patrols by police agency	1,512
• Patrols by non-police agency	143
• Planting new vegetation	400
• Clearing of drainage channels and culverts	101
• Surface maintenance of parking areas	215
• Maintenance of lighting	35
Annual operational/maintenance cost per mile	11,747

Table 8.2 Benefit and cost components of investments in cycling facility
(2012–2042)

Benefit of cycling improvement (present value)	Value ($)
Fuel savings	3,659,677
Parking cost savings	10,228,149
Reduced health care costs	6,494,875
Reduced mortality cost	20,163,000
Reduced accident costs	767,111
Congestion reduction	3,068,445
Vehicle emissions reduction	153,422
Noise reduction	460,267
TOTAL BENEFIT	51,325,945
Costs of cycling improvement (present value)	
Capital costs	19,500,000
Maintenance costs	1,283,349
Tax-cost factor, 20% of budget costs	4,156,670
User costs for bicycle purchase and maintenance	5,813,875
TOTAL COST	30,753,894
Net benefit/cost ratio	0.67

Source: Treatment of the Economic Value of a Statistical Life in Departmental Analyses—2011
Interim Adjustment, U.S. Department of Transportation, updated: October 24, 2012; National
Vital Statistics Reports, Vol. 61, No. 6, October 10, 2012.

Table 8.3 Health economic assessment tool for bicycling (HEAT)

HEAT model inputs	Value
How many people are cycling	110
Average time cycled per person	30 minutes
Average distance cycled per person	6 miles
Average bicycle trips per year	250
Deaths per 100,000 population	799.5 (year 2011)
Value of statistical life	$6.2 million
Time period over which benefits are calculated	30
Costs to include a benefit–cost ratio in the HEAT calculation	No
Mortality rate for the U.S. working-age population	0.0021
Discount rate to apply to future benefits	5%
HEAT model outputs	Unit (USD, 2012)
Value of statistical life applied	6.2 million
Annual benefit of this level of cycling, per year	1,312,000
Total benefits accumulated over 30 years	39,349,000
Current value of the average annual benefit, averaged across 30 years	672,000
Current value of the total benefits accumulated over 30 years	20,163,000

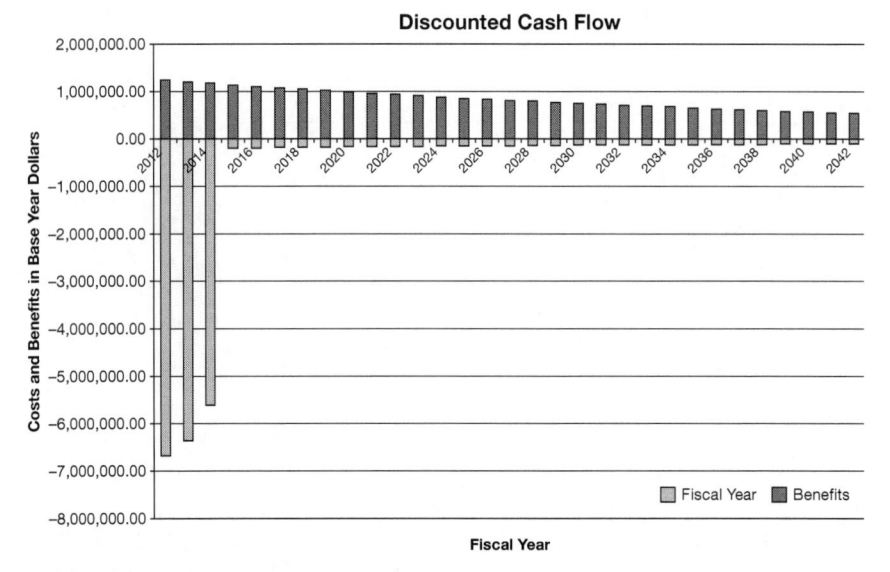

Figure 8.4 Discounted cash flow, where the discount rate equals 5 percent and the annual growth rate mode share equals 2 percent.

economics of a proposed project utilizing multiple tools in the capital budgeting process, e.g., discounted cash flow (DCF) techniques that focus on the evaluation of cash flows taking into account the time value of money. In this way, the model is able to measure discounted cash flow in different periods to have a common basis of comparison (Figure 8.4).

Next, a sensitivity analysis was carried out, in which the effect of annual growth rate of mode share on the net benefit/cost ratios and on the discounted payback period of the project was examined in detail.

Sensitivity analysis of key variables affecting cycling facility

In order to compare the effects of different variables on economic results, sensitivity analysis (or "what if" analysis) was used in this study to emphasize areas where improvements to cycling facility may have the greatest economic impacts and performance. Sensitivity analysis highlighted that the greatest gains to be realized in improving both direct and indirect benefit were those associated with increasing the facility capacity, increasing the mode share rate of cycling, and decreasing the capital cost of the cycling facility.

In the 2000 Census, there were 29,797 city residents in the City of Wilmington who commuted to work. Less than 0.20 percent of commuters listed their primary means of travel via bicycle. In the 2011 American Community Survey, 0.38 percent of commuters listed cycling as their primary mode for travel to work. In contrast, other cities in New Castle County, such as Newark, had 4.7 percent of commuters list cycling as their primary mode for travel to work. In the major U.S. cities studied, the share of commuters by bicycle is 0.8 percent (Steele 2010, 33). Portland retains the highest share of workers commuting by bicycle (6.4 percent) for U.S. cities, with growth in bicycling of 10 percent annually. Portland's long-term goal is to increase bicycling to a mode share of over 25 percent (Gotschi 2011).

This chapter uses data collected from New Castle County, Delaware, to create an analysis framework for estimating the net benefit/cost ratio of improvements to the cycling environment. These NBCRs are conservative, as they do not take into account benefits related to improved livability and economic activity. These benefits are important but cannot be easily measured. It is noteworthy that the analysis demonstrates that a conservative evaluation of benefits based primarily on direct benefits to the users of the cycling facilities can justify possible improvements in bicycle facility selection.

Within the scope of this chapter, the effect of annual growth rate on the net benefit/cost ratios and levels of cycling was studied by means of a sensitivity analysis of cycling projects. The results indicate strong economic viability with positive benefit/cost ratios. Moreover, an increase of up to 10 percent in the mode share increases the net benefit/cost ratio to 1.92.

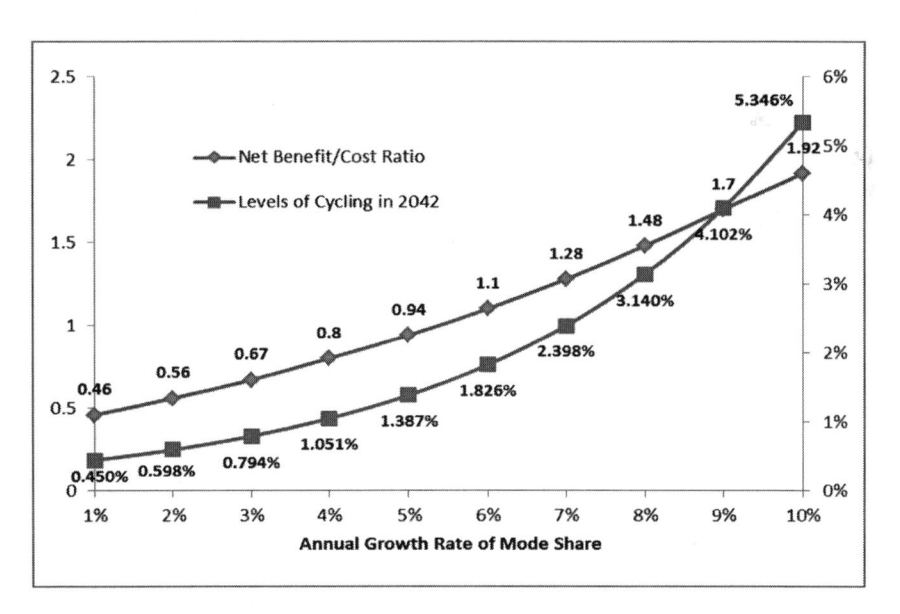

Figure 8.5 Net benefit/cost ratios and levels of cycling at different annual growth rates.

Therefore, the first conclusion that can be drawn is cost–benefit analyses of cycling infrastructure generally produce positive net benefit/cost ratios (see Figure 8.5). However, the question that emerges is how long it takes for an investment to pay for itself, in other words, how rapidly the decision maker can expect confirmation that they have made a good choice. Hence, a second sensitivity analysis is discussed concerning discounted payback period.

As a supplement to profitability measurements, the discounted payback period, where savings are discounted to their present worth before equating to the initial expenditure, is reached during the study period (Figure 8.6). The longer a project's payback period is, the greater the uncertainty or risk of future returns. As shown in Figure 8.6, the discounted payback year lies between 2024 and 2034, i.e., the project has been fully paid back in 2034 when the mode share of cycling increases 2 percent per year or 2024 if the mode share increases 10 percent per year.

It is important to realize that the discount rate is a crucial parameter in determining whether the project will be profitable. In the public domain, discount rates of from 2 to 10 percent are commonly used (Morrison 1998). Figure 8.7 shows the variation in the net benefit to cost ratios under discount rates between 0 percent and 10 percent. The net benefit to cost ratios decrease from 1.8 to 0.1 using a 0-percent and 10-percent discount rate,

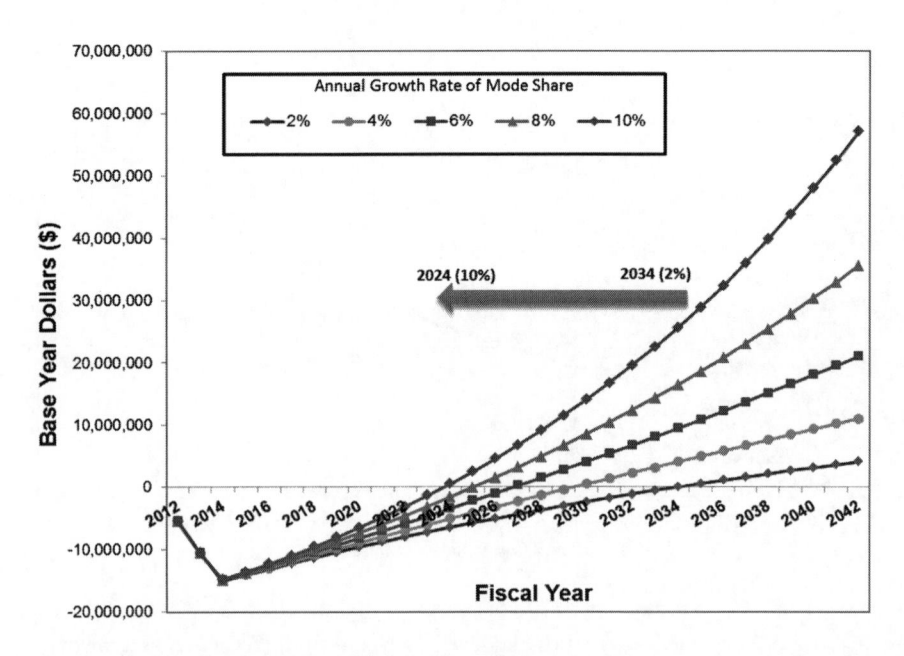

Figure 8.6 Estimated discounted payback period.

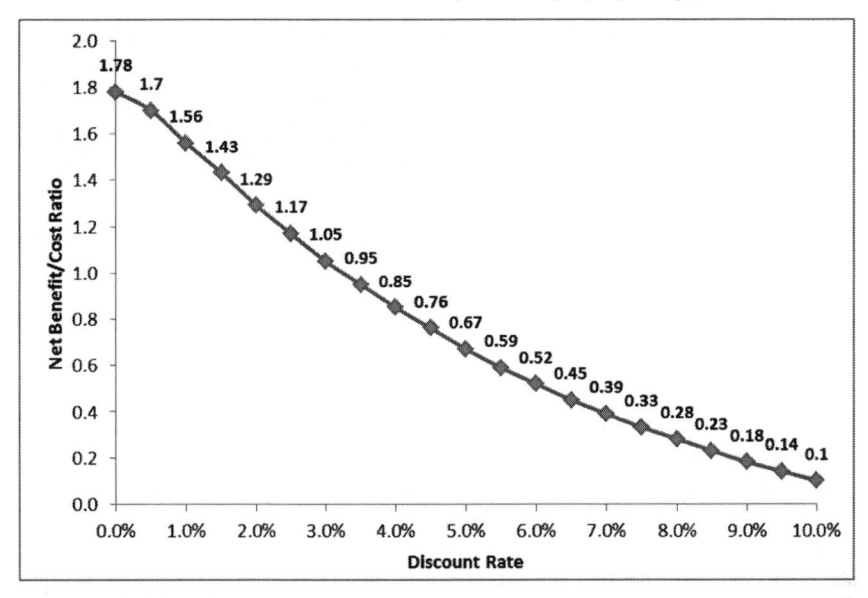

Figure 8.7 Net benefit/cost ratios under various discount rates.

respectively. This is a substantial change, but even with a 10-percent discount rate, the project is still profitable (net B/C ratio >0).

In general, the results of the case study analysis show that implementation of the cycling project will help achieve transport and land use planning objectives including reduced traffic congestion, road and parking cost savings, reduced crash risks, pollution emission reductions, and improved public health. In addition, successful implementation of the larger East Coast Greenway project through Delaware will be an enhancing amenity to the tourist-based Delaware economy. The results of this case study application also serve as a model for how to begin to integrate cycling and its benefits and costs into the general transportation planning process from a Metropolitan Planning Organization's (MPO's) perspective.

Conclusions

One of the most widely used economic evaluation methods in transportation investment is cost–benefit analysis (CBA) which yields estimates of costs and benefits of different types of transportation facilities. It is worth noting that cost–benefit analysis determines the most cost-effective solution, not simply the least-cost solution. In this chapter, a general framework for identifying the effects of adding cycling facilities into an existing road infrastructure was provided to quantify or value the costs and benefits relative to bicycle facilities with CBA methodology that can be applied by transportation planners.

Many different aspects such as traffic congestion, increased travel times, and environmental degradation are already well covered in most CBA studies of transport interventions. Yet too often these do not take account of the wide variety of benefits that arise from the development of infrastructures for cyclists. Our proposed CBA provides urban planners, transportation specialists, and public health advocates with the information necessary to evaluate investments in bicycling. The analysis provides the estimated costs of different bicycle investment options and the benefits to be derived from each feasible alternative. While this chapter focuses primarily on cycling, its findings may also be easily applied to other active modes, such as walking, skateboarding, scootering, and manual wheelchairs.

It is hoped that these results will provide quantitative techniques to assist transportation planners in better understanding and interpreting many of the interrelated processes, the health and environmental benefits from improved cycling facility, and the economic constraints faced by transportation decision makers. Elements of several methodologies were integrated and applied to estimate the costs and the monetary value of the health benefits of active transport modes that could be incorporated into the existing economic evaluation methods. The cost–benefit analysis spreadsheet can be used as part of the Compendium of Supporting Materials. The main limitations are that no benefits related to improved livability and economic activity are considered, and no consideration of walking is made. However, these findings justify the use of the framework in bicycle facility selection or analysis of the characteristics of an existing bicycle facility.

As aforementioned, certain conservative assumptions are necessary in order to improve the potential accuracy of the CBA. Thus, the cost to cyclists due to increased ventilation rate when exposed to air pollution is not included in our model, which can be identified as a potential limitation affecting the validity of the results. For this reason, it may be worth accounting for the effects of air pollution exposure for people using active modes compared with travel by car, and offer transportation planners strategies used to develop non-motorized facility networks that minimize exposure to air pollution. In addition, in order to quantify the benefits that could be realized from increased levels of active transport modes, it would be necessary to investigate the relationships between physical activity and improved livability, increased economic activity, and increased productivity.

Acknowledgments

This work was partially supported by the Delaware Center for Transportation (DCT). The authors express their sincere gratitude to Delaware Department of Transportation (DelDOT) and WILMAPCO Council for providing their extensive data sources.

Notes

1. Some DOTs (such as ODOT) use 4–5 percent, the White House Office of Management and Budget (OMB) use a relatively high central rate of 7 percent, other recommendations suggest 3 percent within 30 years, 2.5 percent between year 30 and 50, and 2 percent~2.5 percent between year 50 and 100.
2. A 20 percent tax cost factor is used to account for the costs to society due to financing that is supported through taxation.

References

Bike Forums. 2005. "How Much Do You Spend on Commuter Biking per Year?" Retrieved Dec. 2, 2012, from www.bikeforums.net/showthread.php/107482-Commuters-How-much-do-you-spend-per-year.

Cavill, Nick, Sonja Kahlmeier, Harry Rutter, Francesca Racioppi, and Pekka Oja. 2007. *Economic Assessment of Transport Infrastructure and Policies. Methodological Guidance on the Economic Appraisal of Health Effects Related to Walking and Cycling.* Copenhagen: World Health Organization, Europe.

Cavill, Nick, Sonja Kahlmeier, Harry Rutter, Francesca Racioppi, and Pekka Oja. 2008. "Economic Analyses of Transport Infrastructure and Policies Including Health Effects Related to Cycling and Walking: A Systematic Review." *Transport Policy* 15(5): 291–304.

Colditz, Graham A. 1999. "Economic Costs of Obesity and Inactivity." *Medicine and Science in Sports and Exercise* 31(11 Suppl): S663–667.

Garrett, Nancy, Michelle Brasure, Kathryn Schmitz, Monica Schultz, and Michael Huber. 2004. "Physical Inactivity: Direct Cost to a Health Plan." *American Journal of Preventive Medicine* 27(4): 304–309.

Gotschi, Thomas. 2011. "Costs and Benefits of Bicycling Investments in Portland, Oregon." *Journal of Physical Activity and Health* 8(1): 49–58.

Haskell, William, I-Min Lee, Russell R. Pate, Kenneth E. Powell, Steven N. Blair, Barry A. Franklin, Caroline A. Macera, Gregory W. Heath, Paul D. Thompson, and Adrian Bauman. 2007. "Physical Activity and Public Health: Updated Recommendation for Adults from the American College of Sports Medicine and the American Heart Association." *Medicine and Science in Sports and Exercise* 39(8): 1423.

Korve, Matthew and Debbie Niemeier. 2002. "Benefit–Cost Analysis of Added Bicycle Phase at Existing Signalized Intersection." *Journal of Transportation Engineering* 128(1): 40–48.

Krizek, Kevin J. 2007. "Estimating the Economic Benefits of Bicycling and Bicycle Facilities: An Interpretive Review and Proposed Methods." In *Essays on Transport Economics*, edited by Pablo Coto-Millán and Vicente Inglada, 219–248. Heidelberg, Germany: Physica-Verlag HD.

Krizek, Kevin, Gary Barnes, Gavin Poindexter, Paul Mogush, and Kristin Thompson. 2006. *Guidelines for Analysis of Investments in Bicycle Facilities.* Washington, D.C.: Transportation Research Board.

Litman, Todd. 2004. "Quantifying the Benefits of Nonmotorized Transportation for Achieving Mobility Management Objectives." Victoria BC, Canada: Victoria Transport Policy Institute.

Lusk, Anne, Patrick Morency, Luis Miranda-Moreno, Walter Willett, and Jack Dennerlein. 2013. "Bicycle Guidelines and Crash Rates on Cycle Tracks in the United States." *American Journal of Public Health* 103(7): 1240–1248.

Marshall, Julian, Michael Brauer, and Lawrence Frank. 2009. "Healthy Neighborhoods: Walkability and Air Pollution." *Environmental Health Perspectives* 117(11): 1752.

Meschik, Michael. 2012. "Reshaping City Traffic Towards Sustainability Why Transport Policy Should Favor the Bicycle Instead of Car Traffic." *Procedia-Social and Behavioral Sciences* 48: 495–504.

Moritz, William. 1997. "Survey of North American Bicycle Commuters: Design and Aggregate Results." *Journal of the Transportation Research Board* 1578(1): 91–101.

Morrison, Edward. 1998. "Judicial Review of Discount Rates Used in Regulatory Cost-Benefit Analysis." *The University of Chicago Law Review*, 1333–1369.

Olabarria, Marta, Katherine Pérez, Elena Santamariña-Rubio, Ana Novoa, and Francesca Racioppi. 2012. "Health Impact of Motorised Trips That Could Be Replaced by Walking." *The European Journal of Public Health* 23(2): 217–222.

Pratt, Michael, Caroline Macera, and Guijing Wang. 2000. "Higher Direct Medical Costs Associated with Physical Inactivity." *The Physician and Sports Medicine* 28(1): 63–70.

Pucher, John and Ralph Buehler. 2006. "Why Canadians Cycle More Than Americans: A Comparative Analysis of Bicycling Trends and Policies. *Transport Policy* 13(3): 265–279.

Rails-to-Trails Conservancy Northeast Regional Office. 2005. *Rail-Trail Maintenance and Operation, Ensuring the Future of Your Rail Trail—A Survey of 100 Rail-Trails.* Camp Hill, PA: Rails-to-Trails Conservancy Northeast Regional Office.

Rojas-Rueda, David, Audrey de Nazelle, Marko Tainio, and Mark Nieuwenhuijsen. 2011. "The Health Risks and Benefits of Cycling in Urban Environments Compared with Car Use: Health Impact Assessment Study." *British Medical Journal* 343: d4521.

Rutter, Harry. 2007. *Health Economic Assessment Tool for Cycling (HEAT for Cycling).* Copenhagen: World Health Organization Regional Office.

Rutter, Harry, Nick Cavill, Francesca Racioppi, Hywell Dinsdale, Pekka Oja, and Sonja Kahlmeier. 2013. "Economic Impact of Reduced Mortality Due to Increased Cycling." *American Journal of Preventive Medicine* 44(1): 89–92.

Salensminde, Kjartan. 2004. "Cost–Benefit Analyses of Walking and Cycling Track Networks Taking into Account Insecurity, Health Effects and External Costs of Motorized Traffic." *Transportation Research Part A: Policy and Practice* 38(8): 593–606.

Schrank, David, Bill Eisele, and Tim Lomax. 2012. *TTI's 2012 Urban Mobility Report.* College Station, TX: Texas A&M Transportation Institute.

Steele, Kristen. 2010. *Bicycling and Walking in the United States: 2010 Benchmarking Report.* Washington, D.C.: Alliance for Biking and Walking.

Taylor, Katherine. 2009. *Utilitarian Cycling: Investigating Latent Demand in Christchurch.* Master's thesis, University of Canterbury.

Turner, Shawn, Aaron Hottenstein, and Gordon Shunk. 1997. *Bicycle and Pedestrian Travel Demand Forecasting: Literature Review.* College Station, TX: Texas A&M Transportation Institute.

U.S. Department of Labor Bureau of Labor Statistics. 2013. *Material and Supply Inputs to Highway and Street Construction, Series ID NDUBHWY--BHWY--.* Retrieved Feb. 10, 2013, from http://data.bls.gov/timeseries/NDUBHWY--BHWY--.

U.S. Department of Transportation. 1999. *Guidebook on Methods to Estimate Non-Motorized Travel*. McLean, VA: U.S. Department of Transportation.

Wang, Guijing, Mike Pratt, Caroline Macera, and Zhi-Jie Zheng. 2004. "Physical Activity, Cardiovascular Disease, and Medical Expenditures in US Adults." *Annals of Behavioral Medicine* 28(2): 88–94.

Wilkinson, Bill. 1997. "Nonmotorized Transportation: The Forgotten Modes." *The ANNALS of the American Academy of Political and Social Science* 87–93.

Wisconsin Department of Transportation (WDOT). 2004. *Wisconsin Bicycle Facility Design Handbook*. Madison, WI: WDOT.

World Health Organization (WHO). 2010. *Global Recommendations on Physical Activity for Health*. Geneva, Switzerland: WHO Press.

9 Secure investment for active transport

Willingness to pay for secured bicycle parking in Montreal, Canada

Dea van Lierop, Brian H.Y. Lee, and Ahmed M. El-Geneidy

Introduction

Active transportation research often focuses on the environmental, economic, health, and social benefits of walking and cycling (Dill 2009; Gordon-Larsen, Nelson, and Beam 2005). While there is much literature about how cyclists use and experience bicycle lanes, boulevards, and paths, few studies evaluate cyclists' perceptions about the security and availability of bicycle parking facilities, especially paid bicycle parking. Much like motorized vehicles, bicycles are more often being kept in parking or storage facilities than cyclists are using them. While the development of cycling networks deserves continued attention, the study of bicycle parking must not be left behind as cities continue to promote active transportation.

Previous studies found that fear of theft and bicycle vandalism discourages bicycle usage among some cyclists (Bachand-Marleau, Lee, and El-Geneidy 2011; Krizek 2006; Schneider 2013; van Lierop, Grimsrud, and El-Geneidy 2015). By providing secured bicycle parking (SBP) for a cost, cyclists can be encouraged to cycle. However, pricing SBP has to be planned and executed carefully to function as an incentive and not a deterrent. The present study recognizes this problem and aims to understand whether users are willing to pay for SBP by examining the following research questions: 1) Are users willing to incur some of the extra cost of improving bicycle parking infrastructure? 2) Of those willing to pay, what are their common characteristics? and, 3) Is there a distinction between those who are willing to pay and those who are able to pay? The data used here are from an online survey conducted in Montreal, Canada, designed to better understand concerns regarding bicycle theft. Although the survey includes information about travel and parking behavior, as well as cyclists' theft histories, this research primarily uses the socio-demographic data and information about participants' willingness to pay (WTP) for SBP to analyze the above-mentioned research questions. While this research is specific to the Montreal region, transportation professionals in other cities can benefit from these findings, which provide insight into the amount cyclists are willing to pay per day for SBP.

Bicycle parking

Transport Canada, the department within the Government of Canada responsible for developing transportation policies and programs (Transport Canada 2010), recognizes that providing SBP is necessary to promote bicycle use. It identifies two types of parking required by cyclists, short-term and long-term, and distinguishes them by design and level of security. Short-term parking can encourage individuals to use bicycles for utilitarian trips (e.g., shopping, running errands); it is most frequently free of charge, located in highly visible outdoor locations, and used by the general public. Short-term parking generally has limited protection from weather, vandalism, and theft. On the other hand, long-term parking can promote bicycle use for commuting since commuter cyclists often need a place to store their bicycles for long periods of time. Some cyclists integrate transit into their trips, thereby requiring long-term SBP at transit nodes, rather than at final destinations. Long-term parking is often made up of bicycle racks in partially or fully enclosed areas or in the form of lockers that enclose each bicycle individually. This type of parking can be located either indoors or outdoors and frequently has higher levels of weather protection and security against vandalism and theft. Some of these facilities charge a fee for usage and are commonly designed for exclusive use by paying cyclists. These facilities are available on a pay-per-use basis or assigned for long-term rentals, including weekly or monthly (Transport Canada 2010).

In Canada, several examples of paid long-term SBP exist. Toronto's Union Station and Victoria Park Bicycle Stations, for example, charge CA$2.15/day, or CA$64.57 for four months plus a one-time CA$26.91 membership fee (City of Toronto 2013). In Montreal, Concordia University's SBP Facility charges staff and students CA$30–$40 a semester (Sustainable Concordia 2013). Metro Vancouver's transportation authority, Translink, provides bicycle lockers at transit interchanges for CA$30 for three months (TransLink 2013). Though no counterpart currently exists in Canada, the US-based consulting, management, and development firm Bikestation has engaged in several public–private partnerships to facilitate the development of SBP facilities (Bikestation 2013). Bikestation charges a US$20.00 annual membership fee plus US$2.00/day for casual users, or a US$96.00 annual fee. Since the installation of bicycle lockers in many cities, the service has become overwhelmingly popular, thereby creating waitlists. Although paid bicycle parking is only sparsely available throughout North America, it is beginning to become more popular in regions where bicycle use is increasing.

This chapter aims to identify and understand the factors that contribute to cyclists' WTP for long-term SBP facilities in Montreal, Canada. It follows the framework of earlier studies that aim to assess users' WTP for a non-market good by using the stated preference contingent valuation method. This method provides quantitative measures to assess the financial value representative of theft-preventing bicycle infrastructure. Since WTP for SBP is a relatively unexplored area of research, the related literature on the contingent valuation (CV)/WTP method, parking pricing strategy, and users' WTP for improved transportation infrastructure is discussed below.

Evaluating cyclists' WTP for SBP

Contingent valuation/willingness to pay method

The CV/WTP method asks individuals to price a service, and uses the stated prices to determine the value of a non-market good. The method is used in the absence of a price for a good and has been tested in many disciplines over the last two decades. It was initially developed in the environmental and public health fields, but has recently been utilized in crime control and justice studies (Cohen et al. 2004; Piquero, Cohen, and Piquero 2011). Like most methods, CV/WTP has strengths and weaknesses. According to Piquero, Cohen, and Piquero (2011), it accurately estimates an individual's attitude toward the perceived price of a good and is useful to place economic value on something that has not previously been assigned a monetary price. Yet, without understanding what the respondent believes to be the cost of the service, it is difficult to determine on what a respondent's stated price is based (Piquero, Cohen, and Piquero 2011). In this study, when determining the appropriate price of SBP in Montreal, individuals are able to state the amount that they would be willing to pay without having been given any indication about how much the costs of SBP would be. A problem with CV/WTP is that individually stated prices may not at all reflect actual costs.

Another issue with this method is that the stated price is not certain to accurately reflect the dollar amount individuals will pay for a service when it becomes available. Cohen (2010) calls this a "hypothetical bias" because the hypothetical dollar value is not always in accordance with the actual dollar value. Cohen (2010) claims that a caution should be made with regard to participants' likeliness to state what they believe is the socially appropriate amount of dollars they are willing to spend, rather than a purely personally evaluated amount (Kahneman et al. 1993). Another common objection to assessing WTP is that it fails to account for ability to pay. For lower income groups, low ability to pay often results in low reported WTP, thereby leading to a greater provision of non-market goods, such as SBP, to higher income groups (SafetyNet 2009).

Charging a fee for bicycle parking is a relatively recent phenomenon; in comparison, paid car parking was first introduced in Oklahoma in 1935 (U.S. Department of Transportation 2012). Manville and Shoup (2005) state that "most cars are parked most of the time, and both auto use and auto ownership are easier if a car can be cheaply and reliably stored when it is not being driven" (233). The same can be said of bicycles, and thus, optimizing the security, design, and availability of both car and bicycle parking facilities deserves attention because it is in these locations that cars and bicycles spend most of their time.

Whereas fees for paid car parking can be set high to function as a negative incentive for driving to certain locations, fees for SBP should not be a disincentive for using a bicycle or a motivation to vacate spots quickly, as with car parking (Shoup 2006). Rather, SBP should be an incentive for bicycle use due to increased levels of security. Finding the right balance between cost and incentive is therefore essential.

Transportation infrastructure improvements

One way to ensure that the pricing of SBP is not too high is to use CV/WTP to determine how much users are willing to pay. Previously, this method has been used to assess WTP for other transport infrastructure, such as the study by Anastasiadou et al. (2009) that determined the demand and appropriate fee for new car parking facilities before construction. Whereas past studies determined parking fees by estimating price-elasticity curves and comparing alternative services, these authors claimed that parking fees should instead be determined based on three principles: the fee should reflect service quality, the economic viability and security of the project, and the demand and needs of the population, especially during peak hours.

Outside of the parking realm, researchers have assessed CV/WTP for other travel-related preferences. For example, dell'Olio et al. (2011) used a stated choice survey to construct logit models to measure individuals' WTP for transfer time, information, and services at transport interchanges. Additionally, Jou et al. (2012) determined freeway drivers' WTP for a distance-based toll, and O'Garra et al. (2007) compared WTP for pollution-reducing hydrogen buses across four cities. More recently, Russo, van Ommeren, and Rietveld (2012) determined university workers' WTP for commuting time.

Although these abovementioned studies are not specifically about pricing bicycle parking, they contain useful information that can help better understand cyclists' WTP for SBP. Particularly, the literature on WTP for transport infrastructure is useful when used in combination with studies that have analyzed how the design, availability, and geographic location of bicycle parking influence ridership. For example, Papon et al. (2011)

surveyed cyclists to determine the most optimal location for SBP and found that most cyclists prefer secured parking near rail stations, and expect it to be free of charge and available 24 hours a day. These authors noted that WTP for SBP is an area of research that requires further attention. Accordingly, due to a lack of literature on cyclists' WTP for SBP, this study sets out to assess whether or not cyclists would be willing to incur some of the costs of SBP in Montreal, Canada.

Study context

The proportion of commuters travelling by bicycle in the Montreal region is 1.7 percent of all trips, which is similar to the cycling mode share in other major Canadian cities such as Vancouver (Transport Canada 2011). The City of Montreal's 2008 Transportation Plan aims to increase the cycling mode share, not only by expanding the bicycle network, but also by increasing the number of parking facilities by 500 percent (Division du Développement des Transports 2008).

In addition to increasing cycling, bicycle parking expansion is intended to reduce bicycle-related crime. According to the city's police department, approximately 2,500 bicycles are reported stolen every year but this number likely represents a small portion of all thefts since many people do not report them (Tremblay and Letendre 2011). A Montreal bicycle theft committee estimated the actual theft numbers to be more likely between 15,000 and 30,000 in 2011 (Riga 2012).

Data and methodology

The data used here were compiled from a bilingual (English/French) online survey on bicycle theft that was conducted in the Montreal region. A variety of measures were taken to allow for broad exposure and reduce sample bias normally associated with online surveys. As recommended by Dillman, Smyth, and Christian (2009), they included circulation through a combination of email newsletters, mailing lists, newspaper articles in French and English, a radio interview, and a number of social networking platforms.

The survey yielded a total sample of 2,039 individuals over a one-month period in late spring 2012. This is similar to the number of home-based cycling trips recorded in the regional Origin–Destination survey, which samples 5 percent of the region's population (Agence Metropolitaine de Transport 2008). While the survey posited a number of questions relating to bicycle theft, this study uses data only from participants who answered the question, "Would you consider paying for supervised or secured bicycle parking? (i.e., security guard, bicycle locker, bicycle parking garage)." The analysis also uses socio-demographic information from the survey, including participants' age, gender, income, employment status, and household size (see Table 9.1). Respondents who left any of these questions unanswered were removed from the sample. The final sample size used in this study is 1,533 Montreal cyclists, of whom 43 percent are willing to pay for secured parking.

To determine the characteristics associated with whether cyclists are willing to pay for SBP this study uses two binary logit models. The dependent variable for these models is derived from the survey question that asked participants about their WTP for SBP. Model 1 includes all of the study participants and is used to identify the factors that most influence cyclists to be willing to pay for SBP. In the second model, which accounts for the potential discrepancy between WTP and ability to pay, only the sample subset with an annual household income greater than \$60,000 is presented. This threshold captures the closest survey income category to Montreal's median total household income the year prior to the administration of the survey (\$69,150) (Statistics Canada 2012), and approximately half of the participants fall into this group, retaining a useful sample size. The results of models 1 and 2 are very similar, revealing that WTP in this study is not greatly affected by ability to pay (see Table 9.2).

Summary statistics

Nearly half of the cyclists in this study are willing to pay for SBP. The respondents' ages range from 18 to 85. The average age for cyclists who are willing to pay is 39, and the average for those not willing to pay is slightly lower at 36. Women, accounting for 42 percent of the survey, are slightly overrepresented, compared to Origin–Destination survey figures (see Table 9.1). Most of the respondents are employed full-time and have completed at least an undergraduate degree. In accordance with the Origin–Destination survey, the largest groups of participants live in two-person households and have a household annual income of between \$20,000 and \$60,000. Approximately 50 percent of the participants reported that they had been victims of bicycle theft in their lifetime, a finding that resembles previous study outcomes in Montreal (Bachand-Marleau, Lee, and El-Geneidy 2011).

Figure 9.1 displays the reported rates that cyclists are willing to pay for SBP as cumulative percentages; it assumes that those willing to pay higher amounts would also be willing to pay lower amounts (i.e., all would be willing to pay \$0). The highest amount that participants are willing to pay is \$15.00. Less than 1 percent of participants are willing to pay more than \$6.00, accordingly not included in Figure 9.1, but 43 percent are willing to pay at least \$0.50. Ideal payments appear to be simple dollar amounts such \$1.00 or \$2.00. These findings are comparable to existing paid facilities where long-term SBP memberships often average less than \$1.00/day, and casual, or daily, SBP is priced at around \$2.00/day (City of Toronto 2013; Cohen et al. 2004; Dill and McNeil 2013; Sustainable Concordia 2013; TransLink 2013).

Table 9.1 Summary statistics

	2012 Bicycle theft survey			2008 Origin–destination survey (Adult)	
	General	*WTP logit*			
	All survey respondents	*Willing to pay*	*Not willing to pay*	*Bicyclists*	*All*
Gender					
Male	58% (1,037)	63% (416)	55% (479)	65% (1,029)	47% (58,890)
Female	42% (738)	37% (249)	45% (389)	35% (548)	53% (65,563)
Age					
Average age	37	39	36	42	48
18–29	30% (542)	26% (175)	31% (270)	24% (372)	16% (19,750)
30–39	37% (658)	35% (234)	39% (342)	22% (343)	16% (20,182)
40–49	17% (301)	17% (110)	16% (140)	25% (395)	21% (25,929)
50–64	14% (254)	20% (130)	11% (99)	24% (371)	28% (34,983)
65+	2% (41)	2% (16)	2% (17)	6% (96)	19% (23,609)
Household size					
One	21% (369)	20% (131)	21% (182)	22% (346)	15% (18,203)
Two	43% (755)	42% (275)	44% (379)	34% (539)	38% (47,008)
Three	19% (335)	19% (129)	19% (160)	20% (310)	19% (24,121)
Four	12% (213)	13% (83)	12% (102)	17% (270)	19% (23,788)
Five or more	6% (100)	7% (44)	4% (38)	7% (112)	9% (11,333)

Table 9.1 Continued

Occupation					
Employed	71% (1263)	80% (533)	70% (608)	68% (1070)	58% (71,544)
Student	21% (370)	14% (93)	24% (207)	13% (200)	8% (9,872)
Retired	3% (50)	3% (18)	3% (22)	11% (181)	25% (31,057)
Other	6% (100)	3% (21)	4% (31)	8% (126)	10% (11,936)
Income (household)					
<$20,000	14% (245)	9% (59)	16% (143)	15% (186)	12% (10,217)
$20,000—$60,000	36% (618)	29% (192)	40% (346)	46% (588)	44% (38,726)
$60,000—$100,000	26% (450)	31% (204)	26% (225)	26% (334)	28% (24688)
>$100,000	23% (391)	32% (210)	18% (154)	13% (166)	17% (15,009)
N*	1,922	665	868	1,577	124,453 (all modes)

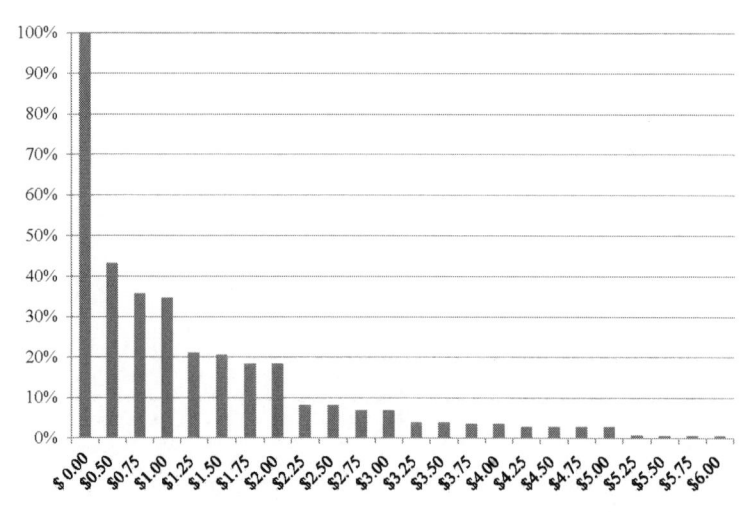

Figure 9.1 Percentage of survey participants' WTP per price category per day.

Further data analysis also showed that cyclists in higher income categories more often were willing to pay for SBP. Similarly, as the price of an individual's bicycle increased, so did his or her WTP for SBP. Similarly, over half of the cyclists who reported owning bicycles worth more than $500, claimed to be willing to pay for parking. These findings substantiate concern that WTP can be influenced by individuals' ability to pay. The differences in participants' WTP and ability to pay are further discussed later in the chapter.

Attitudes towards theft prevention

Before further discussing cyclists' motivations for being willing to pay for SBP, it is helpful to understand that cyclists expressed different theft-preventing attitudes, and that their attitudes were related to their opinions about paying for SBP. More specifically, the results of the survey indicated that there are two contrasting strategies for theft prevention beyond simply locking the bicycles. The first, practiced primarily by owners of expensive bicycles, is to avoid storing a bicycle in open public places. Owners of these bicycles often keep them inside when not in use and are more likely to be willing to pay for SBP. The second strategy, more common with owners of lower value bicycles, is to use electrical tape, anti-theft rust stickers, spray paint, or decoration to make a bicycle less appealing to thieves. Owners of these bicycles are generally not willing to pay for SBP, and alternatively often engage in "fly-parking," the securing of bicycles to street furniture not intended to function as parking.

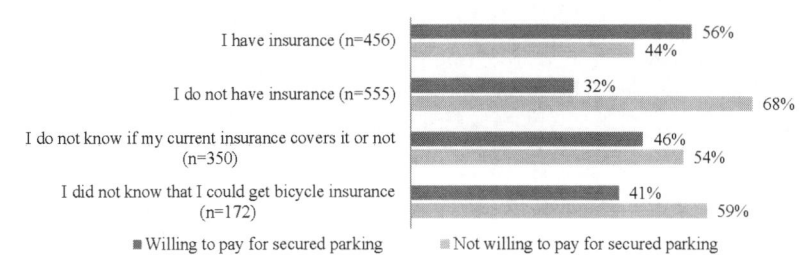

Figure 9.2 Differences in WTP among survey participants who do and do not have insurance for their bicycle(s).

Another way to categorize different cyclists is by whether they have insurance for their bicycle(s). WTP for bicycle parking is clearly reflected in cyclists' WTP for insurance. Figure 9.2 demonstrates that most cyclists who have bicycle insurance are willing to pay for SBP, unlike those in other categories.

Determinants of WTP

To better understand cyclists' WTP for SBP we developed two binary logit models. Model 1 identifies the factors that most influence cyclists' WTP for SBP, and Model 2 accounts for cyclists' ability to pay for SBP, by including only the survey respondents with annual incomes of $60,000 or higher.

To choose the appropriate variables to be included in the models, we developed a correlation matrix. Variables included individuals' habits, choices, and socio-demographic statuses. Many variables pertaining to monetary values such as the use of insurance or the price of the lock used were not included because they were highly correlated with the income variable. Other variables were not included because they did not show significance. Surprisingly, having been a victim of bicycle theft was not statistically associated with a cyclist's WTP for parking.

The models below include information about cyclists' employment status, gender, age, and household income. For example, the variable "gender," although not significant, is included in the models for literature consistency. Cyclists' level of education is not included in the models because it was highly correlated with both employment status and income. The model also includes a continuous variable accounting for the number of years that a cyclist has been a bicycle commuter. Of the attitudinal questions from the survey that considered cyclists' reasons for using a bicycle, only the variable representing the statement, "It is part of my self-identity/culture," was found to be statistically significant and, therefore, retained while other reasons for cycling were removed. Furthermore, the variable describing the cost of the bicycle was included in the model to demonstrate that it is not only how much cyclists earn that affects WTP, but also the amount that cyclists are willing to spend on a bicycle. In addition, how much risk of theft influences a cyclist to use a bicycle was also included, and the relevant literature was consulted to decide which variables should be used.

Model results

The results of the binary logit models are presented in Table 9.2; both possess a reasonable amount of explanatory power (Model 1: AUC = 0.73, Model 2: AUC = 0.72), and both models use WTP for SBP as the dependent variable. Model 1 includes all of the survey participants, and demonstrates that users for whom the risk of bicycle theft influences their decision to cycle are more inclined to be willing to pay for SBP. More specifically, while controlling for other variables, for every one-point increase on a five-point Likert-scale, the odds of an individual being willing to pay for SBP increases by 1.7 times. The odds of being willing to pay for SBP increases with age and is 31 percent lower for students compared to others. With regard to annual household income, cyclists who have an annual income lower than $100,000 have significantly lower odds to be willing to pay for SBP than those with higher incomes. Similarly, the odds of cyclists who own low-value bicycles (under $500) are 50 percent lower compared to cyclists with bicycles valued at over $500. Additionally, as the amount of years that participants have been commuting by bicycle increases, the odds of being willing to pay for parking decreases. This may be due to cyclists' increased level of exposure having led to long-term commuters becoming more aware of theft prevention strategies. Finally, cyclists who report culture/identity as important have a lower odds of being willing to pay for SBP.

Due to the fact that Model 1 demonstrates that income is highly significant, and because several studies have put forth a concern that WTP often does not account for ability to pay (SafetyNet 2009), Model 2 includes only cyclists who have an annual income greater than $60,000 as these users would be more likely to be able to pay compared to those with lower incomes. Unexpectedly, the factors affecting WTP for participants who are most likely also able to pay remains similar to those of the total sample. The results of the models are consistent, suggesting that the significant factors influence cyclists' WTP regardless of their ability to pay.

To further understand the WTP of cyclists who are likely to be able to pay, Figure 9.3 demonstrates the amount these cyclists would be willing to pay for SBP as cumulative percentages, and as in the case of Figure 9.1 assumes that those willing to pay higher amounts would also be willing to pay lower amounts. Of those who are both able and willing to pay for SBP, 84 percent are willing to pay $1.00, and 48 percent are willing to pay $2.00 per day. Both the results of Figures 9.1 and 9.3 clearly demonstrate that most cyclists are not willing to pay rates that exceed $2.00 a day. As mentioned earlier, the results of this study demonstrate that the ideal price appears to be simple dollar amounts, such as $1.00 or $2.00, that incentivize rather than deter people to use a bicycle.

Table 9.2 Binary logit models

Dependent variable: Willingness to pay for parking		MODEL 1 Everyone				MODEL 2 Income > $60,000			
		OR		2.5 %	97.5%	OR		2.5 %	97.5%
(Intercept)		0.66		0.35	1.25	0.35	***	0.20	0.63
To what extent does risk of bicycle theft influence your decision to cycle?	Not at all → Extremely (Likert from 1–5)	1.69	***	1.52	1.86	1.68	***	1.52	1.87
Employment status	Student	0.69	**	0.49	0.97	0.62	***	0.45	0.85
	Not student (reference)	—				—			
Gender:	Male	1.12		0.89	1.41	1.16		0.92	1.45
	Female (reference)	—				—			
Age:	(continuous variable)	1.01	**	1.00	1.02	1.02	***	1.01	1.03
Annual household income:	Less than $20,000	0.50	***	0.32	0.77	NA			
	Between $20,000 and $60,000	0.47	***	0.35	0.64	NA			
	Between $60,000 and $100,000	0.68	**	0.50	0.92	1.07		0.84	1.37
	More than $100,000 (reference)	—				—			
Reason:	Culture	0.63	***	0.50	0.79	0.64	***	0.51	0.81
	Not culture (reference)	—				—			
Years of commuting:	(continuous variable)	0.94	***	0.91	0.97	0.94	***	0.91	0.97
Cost of bicycle:	Low (less than $500)	0.50	***	0.39	0.63	0.47	***	0.37	0.59
	High (more than $500) (reference)	—				—			
Goodness-of-fit measures:		AIC: 1854.6 BIC: 1913.3 AUC: 0.73 †				AIC:1875.3 BIC: 1923.315 AUC: 0.72 †			
		N=1533				N=793			

Signif. codes: 0.001 '***' 0.01 '**'; — = Reference category; NA = Not included in model; † Area under the curve as indicated by receiver operating characteristics (ROC) curve.

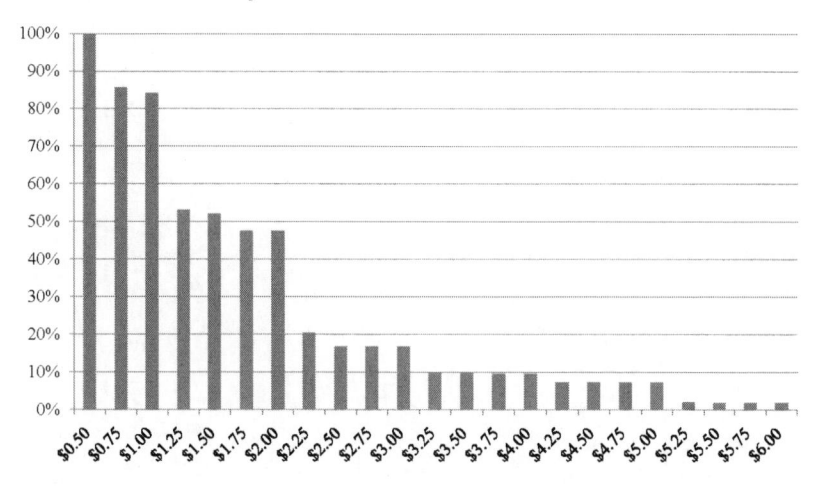

Figure 9.3 Percentage of survey participants with incomes greater than $60,000 and WTP per price category per day.

Discussion and conclusion

Cyclists are likely to become victims of theft or bicycle-related crime over the course of their lives. In our sample, approximately half of the cyclists had a bicycle stolen at least once. The careful development of SBP parking facilities that decrease the chance of theft are likely to encourage individuals to increase their bicycle usage and commute for longer distances.

In addition to implementing regional theft prevention strategies, policy makers and planners should recognize that many cyclists are willing to pay for bicycle security. Although this study provides information about WTP for SBP facilities for a sample of cyclists from Montreal, the findings are also relevant for transportation planners in other regions and this new area of research merits further scholarly attention.

A limitation of this study is that the survey did not ask cyclists who would be willing to pay for SBP how often they would use this service and at which destination(s). Future studies would also benefit from comparing cost estimates to the expected effectiveness of SBP. Another topic that should be addressed is the role of bicycle sharing programs and their relationships with infrastructure investments and cyclists' WTP. Further research should also investigate whether cyclists would rather use a bicycle sharing system and not worry about theft, and relatedly, whether public resources would be better spent on bicycle sharing programs instead of SBP.

Other considerations that city planners and transportation professionals should take into account are the reasons that cyclists would not pay for parking. The responses from the open-ended survey questions confirm that many cyclists are not willing to pay for SBP because they use a bicycle to save money. The statement that "the goal of using a bicycle, among others, is to save money on transportation costs" is representative of the opinions of many survey participants. This finding is also reflected in the summary statistics, which demonstrated that of the people who stated that the low cost of cycling was very or extremely important in their decision to cycle, 61 percent were not willing to pay for SBP. The results of the binary logit model that includes the total sample also confirm that participants with annual household incomes lower than $100,000 are less likely to be willing to pay than those who have higher household incomes. Other reasons that cyclists are not willing to pay for SBP include the concern that the parking would not be located in the places where cyclists would want to go, and that their current bicycle lock was sufficiently secure. One participant stated that "[i]t would take a very long time, if ever, for such services to be located conveniently enough throughout the city. I want to lock my bike close to where I am going." This concern is aligned with the planning norm that land use and transport must be assessed and planned together in order for new transport services (such as SBP) to be optimally enjoyed and used.

Even though many cyclists are not willing to pay (57 percent) for SBP, there is a substantial number who would be interested in increasing bicycle security for a price. Based on the findings of this study, cities will benefit from improving their cycling infrastructure by installing more SBP facilities, since there is a market that exists for these types of facilities and because they are expected to encourage cycling. Cyclists who state that theft risk influences their decision to cycle are more likely to pay for SBP. As theft risk becomes more influential, a cyclist's WTP for SBP increases. If potential cyclists whose mode choice is greatly affected by theft risk can have that risk reduced through SBP, then the amount that they cycle is likely to increase. Therefore, if cities provide more bicycle parking, then bicycle mode share is likely to increase as well. Although the installation of paid SBP is highly recommended, transportation officials should ensure that the pricing of these facilities remains low, at $1.00–$2.00 per day, to ensure that the security provided by paid parking is always an incentive to cycle.

References

Agence Metropolitaine de Transport (AMT). 2008. *Enquête Origine-Destination*. Montréal: Région métropolitaine de Montréal.

Anastasiadou, M., D. Dimitriou, A. Fredianakis, E. Lagoudakis, G. Traxanatzi, and K. Tsagarakis. 2009. "Determining the Parking Fee Using the Contingent Valuation Methodology." *Journal of Urban Planning and Development* 135(3): 116–124.

Bachand-Marleau, Julie, Brian Lee, and Ahmed El-Geneidy. 2011. "Towards a Better Understanding of the Factors Influencing the Likelihood of Using Shared Bicycle Systems and Frequency of Use." *Transportation Research Record* 2314: 66–71.

Bikestation. 2013. *Bikestation: How it Works*. Long Beach, CA: Mobis Transportation Alternatives. http://home.bikestation.com/how-it-works.

City of Toronto. 2013. *Bicycle Parking Stations*. Toronto: City of Toronto. www.toronto.ca/cycling/bicycle-station/.

Cohen, Mark. 2010. "Valuing Crime Control Benefits Using Stated Preference Approaches." In *Cost-Benefit Analysis and Crime Control*, edited by John Roman, Terry Dunworth, and Kevin Marsh, 73–118. Washington, D.C. Urban Institute Press.

Cohen, Mark, Roland Rust, Sara Steen, and Simon Tidd. 2004. "Willingness-to-Pay for Crime Control Programs." *Criminology* 42(1): 89–110.

dell'Olio, Luigi, Angel Ibeas, Patricia Cecín, and Francesco dell'Olio. 2011. "Willingness to Pay for Improving Service Quality in a Multimodal Area." *Transportation Research* Part C 19: 1060–1070.

Dill, Jennifer. 2009. "Bicycling for Transportation and Health: The Role of Infrastructure." *Journal of Public Health Policy* 30(S1): S95–S110.

Dill, Jennifer and Nathan McNeil. 2013. "Four Types of Cyclists? Examining a Typology to Better Understand Bicycle Behavior and Potential." Paper presented at the Transportation Research Board 92nd Annual Meeting, Washington, D.C.

Dillman, Don, Jolene Smyth, and Leah Christian. 2009. *Internet, Mail and Mixed-Mode Surveys: The Tailored Design Method*, 3rd. ed. Hoboken, NJ: John Wiley & Sonson.

Division du Développement des Transports, ed. 2008. *Transportation Plan 2008 – Ville de Montreal*. Montreal, Canada.

Gordon-Larsen, Penny, Melissa Nelson, and Kristin Beam. 2005. "Associations Among Active Transportation, Physical Activity, and Weight Status in Young Adults." *Obesity Research* 13(5): 968–975.

Jou, Rong-Chang, Yu-Chiun Chiou, Ke-Hong Chen, and Hao-I Tan. 2012. "Freeway Drivers' Willingness-to-Pay for a Distance-Based Toll Rate." *Transportation Research* Part A 46 (3): 549–559.

Kahneman, Daniel, Ilana Ritov, Karen Jacowitz, and Paul Grant. 1993. "Stated Willingness to Pay for Public Goods: A Psychological Perspective." *Psychological Science* 4(5): 310–315.

Krizek, Kevin 2006. "Approaches to Valuing Some Bicycle Facilities' Presumed Benefits." *Journal of the American Planning Association* 72(3): 309–320.

Manville, Michael and Donald Shoup. 2005. "Parking, People, and Cities." *Journal of Urban Planning and Development* 131(4): 233–245.

O'Garra, Tanya, Susana Mourato, Lisa Garrity, Patrick Schmidt, Anne Beerenwinkel, Matthias Altmann, David Hart, Cornelia Graesel, and Simon Whitehouse. 2007. "Is the Public Willing to Pay for Hydrogen Buses? A Comparative Study of Preferences in Four Cities." *Energy Policy* 35: 3630–3642.

Papon, Francis, Philippe Assaf, Kely Berezoski, Cristina Osipov, and Edward Santa Maria Davila. 2011. "Intermodal Bicycle Parking Facilities: A Stated Preference Survey." Paper presented at the Transportation Research Board 90th Annual Meeting, Washington, D.C.

Piquero, Nicole, Mark Cohen, and Alex Piquero. 2011. "How Much is the Public Willing to Pay to be Protected from Identity Theft?" *Justice Quarterly* 28(3): 437–459.

Riga, Andy. 2012. "McGill Report: Sturdier Racks the Single Biggest Subject of Concern." *The Gazette*, 3 August.

Russo, Giovanni, Jos van Ommeren, and Piet Rietveld. 2012. "The University's Workers' Willingness to Pay for Commuting." *Transportation* 39(6): 1121–1132.

SafetyNet. 2009. *Cost–Benefit Analysis*. Montreal: Transport: Directorate-General Transport and Energy.

Schneider, Robert J. 2013. "Theory of Routine Mode Choice Decisions: An Operational Framework to Increase Sustainable Transportation." *Transport Policy* 25: 128–137.

Shoup, Donald C. 2006. "Cruising for Parking." *Transport Policy* 13: 479–486.

Statistics Canada. 2012. Median Total Income, by Family Type, by Census Metropolitan Area in All Census Families. Ottawa: Government of Canada.

Sustainable Concordia. 2013. "Secure Bike Parking Facility." Concordia University, Montreal, Quebec. http://sustainable.concordia.ca/working-groups/allego-concordia/projects/secure-bike-parking-facility/.

TransLink. 2013. "Bike Lockers." Vancouver Metro BC TransLink. www.translink.ca/en/Rider-Guide/Bikes-on-Transit/Bike-Lockers.aspx.

Transport Canada. 2010. *Bicycle End-of-Trip Facilities: A Guide for Canadian Municipalities and Employers*. Ottawa: Minister of Transport.

Transport Canada. 2011. *Proportion of Workers Commuting to Work by Car, Truck or Van, by Public Transit, On Foot, or By Bicycle, Census Metropolitan Areas*. Ottawa: Minister of Transport.

Tremblay, A. and N. Letendre. 2011. *Service de Police de La Ville de Montréal*. Montréal.

U.S. Department of Transportation. 2012. *Contemporary Approaches to Parking Pricing: A Primer*. Washington, D.C.: Federal Highway Administration.

van Lierop, Dea, Michael Grimsrud, and Ahmed El-Geneidy. 2015. "Breaking into Bicycle Theft: Insights from Montreal, Canada." *International Journal of Sustainable Transportation* 9(7): 490–501.

10 Site suitability and public participation

A study for a bike-sharing program in a college town

Yuwen Hou and Mônica A. Haddad

Introduction

Concerns about global climate change, energy security, and unstable fuel prices have motivated decision makers worldwide to explore sustainable options for transportation (Schäfer et al. 2009). One strategy supported by transportation planners is bike-sharing programs (BSPs), which ease traffic volume and provide sustainable and green options for urban environments (García-Palomares, Gutiérrez, and Latorre 2012). Users of BSPs can take advantage of biking without the responsibilities of bike purchases, maintenance, and obligations related to parking and storage. Moreover, BSPs incorporate cycling into the public transportation system (Shaheen, Guzman, and Zhang 2010), providing transit users an option that offers mobility and flexibility at a lower cost (TransLink 2008). In the U.S., there are several implemented examples of BSPs, such as Hubway in Boston, MA; Smartbike in Washington, D.C.; and NiceRide in Minneapolis, MN (FHWA 2012). However, according to DeMaio and Gifford (2004), BSPs are not suitable for all American cities. BSPs are most appropriate for "urban areas with more compact downtowns, university campuses, and dense neighborhoods with a concentration of younger people" (DeMaio and Gifford 2004, 11).

As an effort to alleviate negative traffic impacts in surrounding neighborhoods, offer affordable transportation choices, as well as boost health and wellness and reduce infrastructure costs, universities tend to support environmentally friendly and cost–benefit efficient solutions such as sustainable transportation projects, including BSPs (Toor and Havlick 2004, 5). Due to the high number of commuters, university campuses have long been associated with the promotion of sustainable travel demand management (TDM) (Balsas 2003; Bond and Steiner 2006; Dagget and Gutkowski 2007). University campuses are unique built environments that serve as significant trip-attractors for TDM. According to the Association for the Advancement of Sustainability in Higher Education (Tang 2010), in 2010, approximately

90 universities in the United States offered campus BSPs of one type or another. Some programs were exclusively designed for campus, such as Zot Wheels at University of California, Irvine, California, while others have stations all over the community, including the campus, such as the B-cycle Program in Boulder, Colorado.

The location and spatial distribution of bike-sharing stations are key factors to be considered for bike-sharing programs' successful implementation. Thus, the overall question this chapter attempts to answer is: Where are the most suitable locations for bike-sharing stations in a college town? The main objective of this study is to identify potential locations for bike stations throughout the City of Ames, Iowa, including the Iowa State University campus and surrounding neighborhoods. Towards that end, a mixed methods approach is used, incorporating Geographic Information Systems (GIS) and gathering public input through a community survey.

We believe that the methodology presented in this chapter can be useful to provide public bike services and to promote biking as a sustainable transportation mode in the City of Ames, Iowa. Our intention is to develop a practical site selection tool for bike stations that can link academic studies and field practices. Academic studies tend to be more focused on technological models (see, for example García-Palomares, Gutiérrez, and Latorre 2012; Krykewycz et al. 2010; Maurer 2012), while field practices tend to involve public participation more directly through surveying, online interactive mapping, and other instruments. For example, the City of Detroit, Michigan, employed an interactive mapping website to solicit input from the public (Detroit Bicycle Sharing 2013). In our study, we use spatial analysis for the initial site identification and ask potential users to have a voice in the site selection process. By taking into account the perceptions of Ames' population in the spatial analysis, we engage the public into the planning process in a manner that links modeling and public participation.

This chapter is organized as follows. First, we introduce the study area. Second, we present a literature review focusing on BSPs and bike stations. Following that, we present our conceptual framework, including the GIS modeling. The final sections focus on the public participation component and the conclusions and limitations of the study.

The case of Ames, Iowa

The City of Ames is the home of Iowa State University (ISU). It is located in central Iowa and has a population of 59,042 (U.S. Census Bureau 2010). Figure 10.1 presents a map of Ames depicting current and future bike paths. ISU has dedicated continuous efforts towards achieving its aspiration to become a sustainable campus. One example of these efforts is the "Live Green!" Initiative that began in 2008. Another example is the Symposium on Sustainability, held annually on campus. One strategy proposed during the 2009 Symposium on Sustainability was to "provide infrastructure for

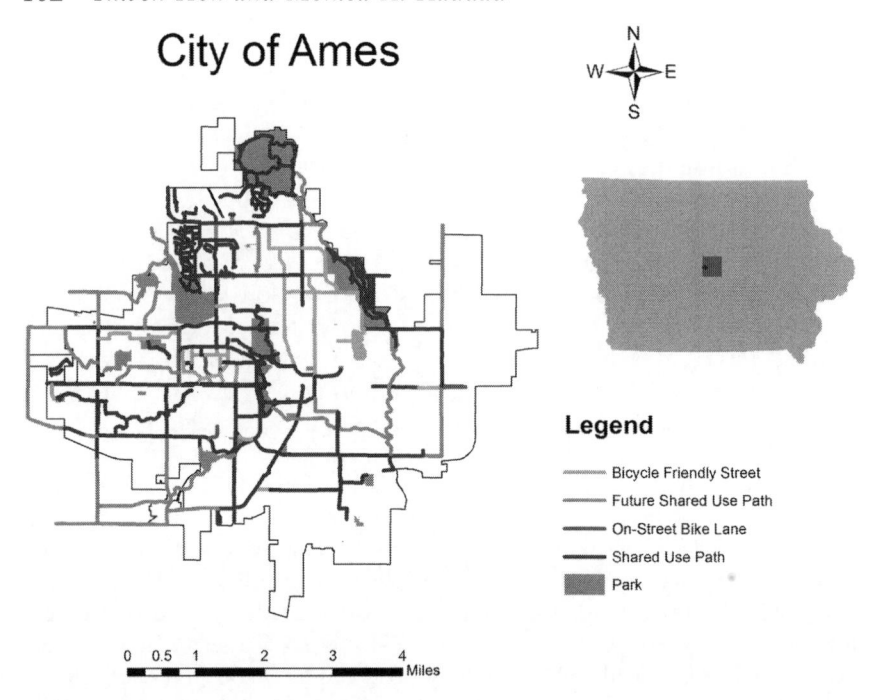

Figure 10.1 Bike paths in Ames.
Source: City of Ames, Iowa.

alternative transportation" (ISU Live Green! 2009). A BSP could contribute significantly to achieving this goal, and should be considered when ISU decision makers seek to expand the current on-campus bike services, such as the existing bike rental service. Furthermore, as enrollment and employment at ISU continue to grow, substantial traffic and parking issues make it increasingly important to introduce strategies such as BSP.

Additionally, the Ames Area Metropolitan Planning Organization 2035 Long-Range Transportation Plan was found to be a significant source of information for our study. Its vision statement declares: "The Ames area future transportation plan delivers innovative and forward-thinking mobility solutions that respond to its unique character as a university community and provides long term sustainability" (City of Ames 2010, 2–1). Implementing a BSP supports this vision.

Literature review

Geographic information systems and bike-sharing stations

Various scholars have applied GIS as part of the methodology to examine bike-sharing programs, especially in site selection studies. Studies that utilized

GIS in Philadelphia, Pennsylvania, Sacramento, California, Milwaukee, Wisconsin, and Madrid, Spain, are examined here. Two important GIS concepts, raster data and Weighted Sum Overlay, should first be introduced in order to better understand the studies described in this section. Raster data is a type of spatial data that consists of a matrix of cells (or pixels) organized into rows and columns (or a grid) where each cell contains a value representing information. To illustrate, an input raster layer representing job density would have a job density value in each cell, organized by location. The Weighted Sum Overlay is an ArcGIS software tool, part of the Spatial Analyst toolbox. In order to use this tool, one has to input raster layers, which are given weights (values), and then an overlay analysis of the layers is performed. The output layer of the Weighted Sum Overlay tool is also a raster. Each cell in the output raster is given a value, which is the sum of all scores of cells located in the same place.

The Krykewycz et al. (2010) study aimed to define the market area for a large-scale BSP in Philadelphia, PA. The authors developed a raster-based GIS analysis and used demographic data, infrastructure data, as well as land use factors to identify the geographic boundaries of the primary and secondary service areas for a new BSP. Variables such as population and job density, locations of tourist attractions and bus stops, and proximity to parks, rail stations, and bike lanes were included. The researchers used "bike-share trip diversion rates" created by peer European cities to estimate daily trips that would be generated by the proposed BSP in order to define different demand scenarios (Krykewycz et al. 2010). This analysis also produced daily usage estimates under different demand scenarios based on low, middle, and high demand.

In another study that applied GIS to define service areas, Maurer (2012) included demographic data and socioeconomic conditions to determine the service areas for the City of Sacramento, CA. Also drawing upon data from an existing BSP in Minneapolis, MN, she created a regression model using monthly rentals as a dependent variable. After Maurer identified the determinants for bike rentals, she then established a bike rental predictive model for Sacramento. A weighted sum raster analysis was also performed to identify suitable locations for a BSP in the recommended service area.

Rybarczyk and Wu (2010) proposed a combined approach that included multi-criteria evaluation and exploratory spatial data analysis (ESDA) for bike facility planning. The MCE analysis integrated supply and demand variables, based on transportation analysis zone level. The ESDA was conducted at the neighborhood level to explain the spatial patterns of bike facility planning in Milwaukee, WI. The authors also used different units of analysis, creating customized analyses for government agencies, planners, and bicycle advocates.

Lastly, García-Palomares, Gutiérrez, and Latorre (2012) carried out a BSP location optimization study for Madrid, Spain. Using GIS, they calculated the distribution of potential trip demand in the study area. Then they applied location-allocation models, using five different scenarios. After examining each scenario, based on different numbers of stations, they concluded that a practical approach to locate stations is to "minimize impedance and maximize station coverage" (García-Palomares, Gutiérrez, and Latorre 2012, 239). In other words, it is better to have stations that encompass larger service areas and are connected with each other, rather than to have more stations that are not connected.

These four examples include common variables used when planning for bike facilities, and more specifically BSPs, reflecting the "4-Ds" of transportation planning: density, diversity, destination, and design. Variables such as employment density, population density, and land use diversity represent the demand side of bike trip generators; destinations associated with recreational purposes, such as parks, often attract biking trips; and lastly, some design features, especially existing facilities, often indicate higher possibilities of biking trips. These previous studies laid the foundation for the choice of variables to be included in our GIS model.

Public participation

Public participation is crucial in transportation planning processes. Transportation agencies should ensure clarity, accessibility, and opportunities for the public to provide input during all stages of transportation planning. Surveying is one of the most effective methods for gathering public input for station site selections. Public participation in the practices of BSPs can occur when the public is invited to examine the possible locations of bike stations.

To illustrate, in the process of developing the Citi Bike program in New York City, practitioners included public input through various approaches such as public meetings and an interactive station planning mapping system (NYC DOT 2013). Another example is the BSP planning process in Detroit, MI. Their web page included a "locate a station" tab to allow the public to be engaged in the process of choosing bike stations (Detroit Bicycle Sharing 2013). Such practices provide opportunities for the public to participate in the planning process for bike stations in their communities.

Conceptual framework

Based on the literature review, we found that when researchers plan for BSPs, models are used to identify locations for bike stations. These models, which are similar to travel demand models, necessitate the imputation of socioeconomic characteristics, land use, neighborhood destinations, and other variables into spatial criteria. Exactly which spatial criteria are needed vary from city to city. On the other hand, when planners need to identify locations for bike stations, they solicit public input as part of their process. Each approach (i.e. researchers' and planners') has its pros and cons. Spatial models usually reveal large-scale physical constraints and also display opportunities such as existing bike lane networks. However, such an approach can be limited, depending on the availability and accuracy of up-to-date spatial data (Ramirez 1996). Including public participation is essential to assure that citizens have their voices heard in the planning process as well as have the opportunity to share local knowledge and ideas. Public participation, however, can be both time-consuming and expensive.

In this study, we used spatial models *and* public participation as a strategy to complement the strengths of both approaches. As displayed in Figure 10.2, by combining the results from GIS modeling and a community survey, we were able to identify appropriate locations for bike stations in a study area.

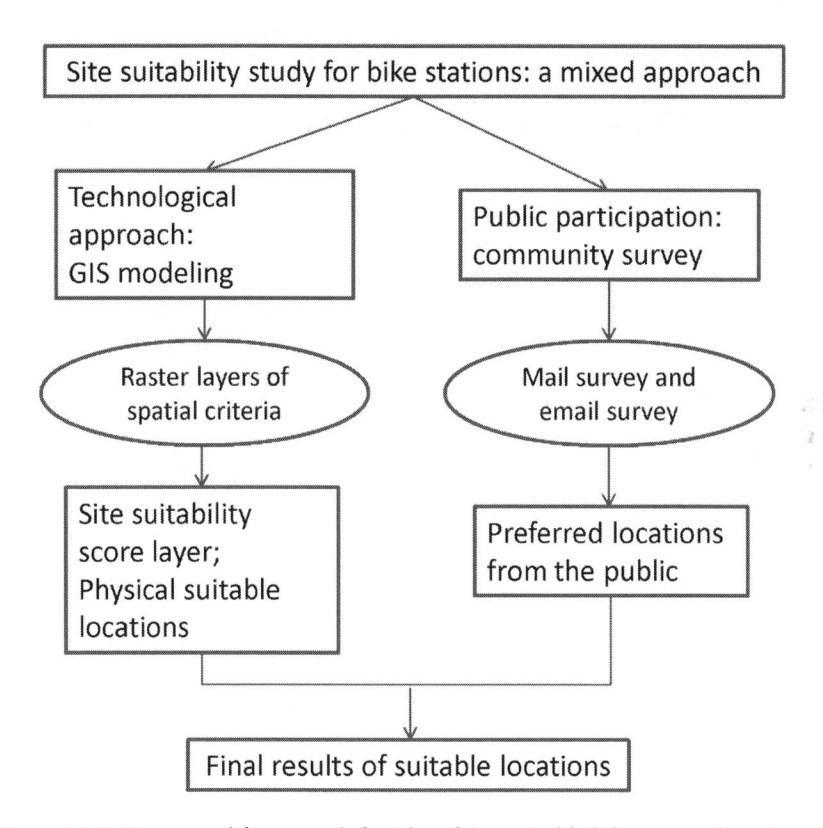

Figure 10.2 Conceptual framework for identifying suitable bike station locations.

GIS modeling

We used the following spatial criteria to identify potential locations for bike-sharing stations in the City of Ames, Iowa:

1. Trip generator—represents the demand side (i.e., potential bikers). Bike stations should be located close to places with higher populations, with higher median household incomes,[1] and with higher numbers of alternative commuters.
2. Trip attractor—represents the supply side (i.e., the bike stations themselves). Bike stations should be located close to attractions, to neighborhood parks, and to places with higher numbers of jobs.
3. Transportation network—represents the transportation infrastructure (i.e., existing bike lanes and bus routes). Bike stations should be located close to bus stops and existing bike lanes.

Attractions include community centers, shopping centers, museums, and civic centers (Maurer 2012). Alternative commuters are defined as the sum number of people who commute by: public transit, taxis, motorcycles, bicycles, walking, and other means. Table 10.1 provides an overview of variables used in this study.

Table 10.2 shows the descriptive statistics of "population," "median household income," "number of jobs," and "number of alternative commuters." Population and job numbers were gathered at the census block spatial scale, having 1,082 blocks with population higher than zero, and 359 blocks with number of jobs higher than zero. Median household income and number of alternative commuters were gathered at the census tract spatial scale, having 13 tracks with median household income higher than zero, and 14 tracts with number of alternative commuters higher than zero. One can observe that the mean for population is 54.5, the mean for number of jobs is 24, the mean for median household income is $48,414, and the mean for number of alternative commuters is 522.

Other spatial layers for the analysis were obtained from the City of Ames, including "attractions," "neighborhood parks," "bus stops," and "bike lanes." We received these spatial layers in vector format. We then transformed them into raster layers with a cell size of 10 by 10 meters. The larger the cell size, the more generalized will be the output of the analysis. To perform the GIS modeling, we defined a common scale to reclassify each raster layer, as presented in Table 10.3. This scale ranges from "very good" to "bad/no data." The reclassification process changes the value in each cell based on a common scale, and defined by the analysts.

After applying the reclassification, all the values of the raster layers were changed. The new cell values ranged from 5 to 0. The reclassification intervals for the variables "population," "median household income," "number of alternative commuters," and "number of jobs" were defined

Table 10.1 Variables for spatial analysis

Definition	Reference	Source	Unit of analysis	Year
Population	Krykewycz et al. 2010, Maurer 2012, García-Palomares et al. 2012	City of Ames	Census block	2010
Median Household Income	Krykewycz et al. 2010, Maurer 2012	Census Bureau, ACS 2010 5-year estimate	Census tract	2010
Number of alternative commuters	Maurer 2012	Census Bureau, ACS 2010 5-year estimate	Census tract	2010
Number of Jobs	Krykewycz et al. 2010, Maurer 2012, García-Palomares et al. 2012, Rybarczyk and Wu 2010	Census Bureau, LEHD database	Census block	2010
Attractions	Krykewycz et al. 2010, Maurer 2012, García-Palomares et al. 2012	City of Ames	Entire study area	Most recent
Distance to neighborhood parks	Krykewycz et al. 2010, Maurer 2012	City of Ames	Entire study area	Most recent
Distance to bus stops	Krykewycz et al. 2010, Maurer 2012	City of Ames	Entire study area	Most recent
Bike lanes	Krykewycz et al. 2010, Maurer 2012	City of Ames	Entire study area	Most recent

Table 10.2 Descriptive statistics of 2010 U.S. Census Bureau variables

Variable	Max	Min	Mean	Standard deviation	Count
Population	2,355	0	54.5	144.9	1,082
Median household income	88,914	20,878	48,414	22,126	13
Number of alternative commuters	1,221	38	522.3	321.8	14
Number of jobs	153	1	24	34.9	359

Source: American Community Survey 2010.

using Jenks (1967) "natural breaks." Jenks' algorithm minimizes each class's average deviation from the mean, while maximizing each class's deviation from the means of other groups, which reduces the variances within classes while maximizing the variance between classes (Jenks 1967). This method emphasizes the differences between created classes and works best with less than seven classes. Distances to parks, bus stops, and other attractions were calculated based on the Euclidean Distance Tool in ArcGIS 10.1. The Euclidean Distance Tool assigns cell values by calculating the distance from each cell to the nearest attraction. Existing "bike lanes" were labeled "very good" only.

Next, all eight raster layers were overlaid using the Weighted Sum Overlay tool, with each layer receiving an equal weight of one (Maurer 2012). The final raster, combining all eight layers, is displayed in Figure 10.3. The legend values, ranging from 31 to 4, are the mathematical sum of the eight input layers, following the scale depicted in Table 10.1. As depicted by the darker area in Figure 10.3, the higher the legend value of a cell, the higher the site suitability score. Areas with higher scores are priority locations when placing stations for bike-sharing.

The cell size for the overlay result layer was set at 10 by 10 meters. However, it is important to observe that the spatial unit for socioeconomic variables, as indicated in Table 10.1, was derived from the Census tract. This resulted in very homogeneous areas, even though the cell sizes were only 10 by 10 meters. As a consequence, the map displayed in Figure 10.3 has some areas with continuous tone because the cells received the same score based on the Census tract's spatial scale. Using the site suitability score map, shown in Figure 10.3, four areas were identified with the highest suitability scores (darker gray), indicating potential areas to locate bike-sharing stations: West Ames residential area, Central ISU campus, Ames Downtown area, and North Ames commercial area.

To assure that our results were consistent, we conducted a sensitivity analysis (SA). Crosetto and Tarantola (2010) discuss the importance of SA in GIS modeling based on the assumption that input factors should be independent. SA techniques provide a basic framework to explore the change

Table 10.3 Reclassification scale of raster layers

Scale	Population	Median household income	Number of jobs	Number of alternative commuters	Distance to parks, bus stops, attractions (m)	Bike lanes
Very good (5)	997–2,355	61,991–88,914	930–2,719	708–1,221	0–200	Existing
Good (4)	521–996	51,735–61,991	389–929	593–707	200–400	NA
Adequate (3)	271–520	37,028–51,735	200–388	384–592	400–600	NA
Poor (2)	150–270	26,010–37,028	86–199	331–383	600–800	NA
Very poor (1)	52–149	21,343–26,009	28–85	141–330	800–1,000	NA
Bad or nodata (0)	<52	<21,343	<28	<140	>1,000	Non existing

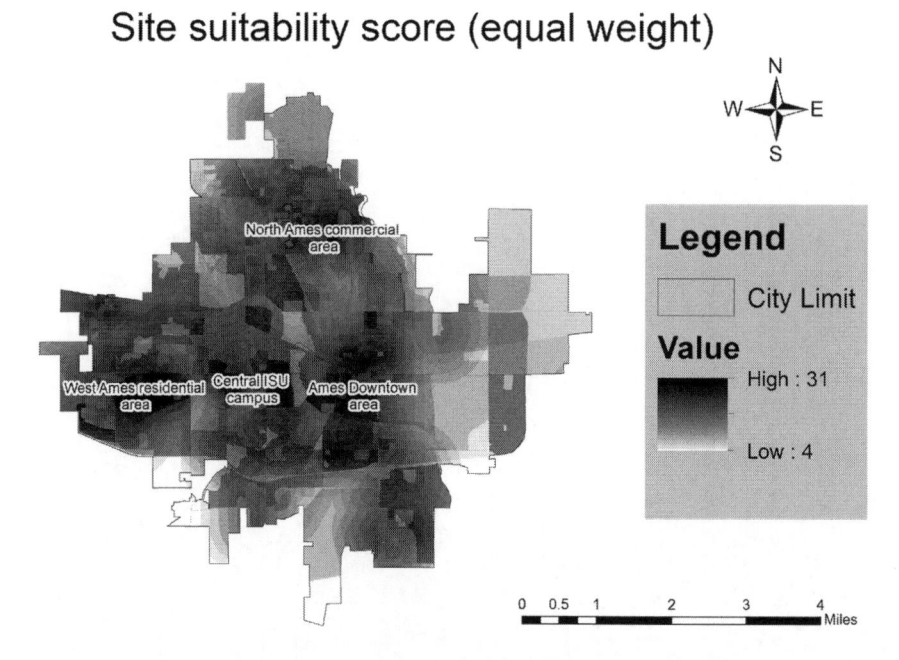

Figure 10.3 Equally weighted sum overlay for the site suitability model.

of outputs when inputs change. In our SA, all inputs were reclassified based on an increment of 5 percent.[2] SA resulted in no significant and proportionate changes to the new outputs, indicating that the site suitability model depicted in Figure 10.3 was robust.

Public participation: the community survey

The next step of our study was to conduct a survey of community residents to identify their preferred locations for bike stations. Having potential users involved in the process of identifying suitable locations allowed their preferences to be combined with GIS-derived results, further improving our site suitability assessment model. Since the residents of the City of Ames as well as ISU students are potential users of a BSP, they together comprised the study population from which the survey sample was selected.

Bromley (2006) reviewed the relationship between campuses and their surrounding communities, focusing on aspects from institutional competition to regional planning priorities. Based on the unique demographic characteristics of campuses and their surrounding communities, i.e., students and residents, we assumed that the perceptions, attitudes, and preferences towards BSPs might contrast. The majority of students live in a campus town only during the years of their education. On the other hand, residents have a vested interest in the cities where they live. Paper surveys were mailed to Ames residents because we did not have access to e-mails, only to their mailing addresses. Whereas, ISU students, both undergraduate and graduate, received e-mails with the link to the survey (via SurveyMonkey) using the same set of questions used for the paper survey.

To define an appropriate sample size, Cochran's formula was used for calculating sample sizes for a large population (Cochran 1977). The sample for Ames residents was defined as follows. The total population of Ames in 2010 was 58,965 (U.S. Census Viewer). With a confidence level set at 95 percent and a desired level of precision set at 5 percent, the sample size was 384. According to the U.S. Census, which has a 2010 population of 58,965 (U.S. Census Viewer), people should be enumerated at a residence if they: "live or stay at the residence most of the time; OR Stayed there on April 1, 2010 and had no permanent place to live; OR Stay at the residence more time than any other place they might live or stay," which means students living on campus as residents for the duration of a school-year should not be counted towards the residency of college towns.

Surveys were printed and mailed out to 500 randomly selected Ames resident respondents in March 2013. The sample for ISU students was defined as follows. The total ISU enrollment in the fall of the academic year 2012–2013 was 31,040. With a confidence level set at 95 percent and a desired level of precision set at 5 percent, the sample size was 379. Surveys were designed using Survey Monkey and sent by email to 2,000 randomly selected undergraduate and graduate students in March 2013. We received 67 valid responses from the residents, representing 17.4 percent of the ideal sample size for Ames residents and a total of 158 responses from the online survey, representing 41.7 percent of the ideal sample size for ISU students. Neither group of response rates reached the ideal sample size. Therefore, the result of surveys cannot be taken to represent the entire population.

The main objective of the survey was to capture where residents and students would like to have bike stations located, based on interests and convenience. The survey had eight questions related to BSPs. In this chapter, we examine in detail the two questions related to the location of bike stations. Other questions included in the survey were aimed at understanding the attitudes, conceptions, and preferences of the Ames public towards biking and bike-sharing programs in general. Table 10.4 presents the survey questions.

Before analyzing the two questions related to bike station locations, we will examine the questions asked in the survey about BSPs in Ames. These questions asked the respondents their knowledge of BSP and whether they perceive a potential need for a BSP in Ames. The responses are displayed in Figure 10.4 a and b. Most of the respondents either "agreed" or "strongly agreed" with the statement that "Ames is a good place for a bike-sharing program," corresponding to approximately 70 percent of Ames residents and 82 percent of ISU student respondents. For the statement "If there was

Table 10.4 Survey questions

1. Do you own a bike?	The respondents were asked to check one answer that applies with their ownership of bikes.
2. How often do you ride a bike in Ames?	The respondents were asked to check one answer that applies with the frequency of riding bikes.
3. Why do you ride a bike in Ames?	The respondents were asked to check all answers that apply with the purpose of riding bikes.
4. Below are some questions about riding preferences. Check the answer that best describes you for each question.	The respondents were asked to check Yes/No/Don't know for descriptions about riding preferences such as on/off-street, available biking preferences.
5. Have you heard of bike sharing before now?	The respondents were asked to check on answer that best fit with their awareness the concept of bike sharing.
6. Please check the box for the top three places that you think would be the best locations for bike sharing station in Ames.	The respondents were asked to choose three out of 12 potential locations for bike station, with a map of Ames showing the locations for reference.
7. Please indicate your level of agreement with the following statements about bike sharing programs.	The respondents were asked to reveal their attitudes towards installing bike sharing in Ames.
8. What physical features do you think bike-sharing stations should be close to?	The respondents were asked to rank six features to indicate their priorities when locating bike stations.

a bike-sharing program in Ames, I would use it," there was a clear difference: around 55 percent of the students strongly agreed or agreed and around 31 percent of the residents strongly agreed or agreed. This indicates that students and residents generally hold a welcoming attitude towards a BSP in Ames; however, more students than residents might be likely to use a BSP. With regard to the statement "I will ride bikes more often if there is a bike-sharing program in Ames," 58 percent of the students strongly agreed or agreed with it; on the other hand, 37 percent of the residents disagreed or strongly disagreed with this statement, and only approximately 34 percent of residents strongly agreed or agreed. Again, students showed higher interest in a BSP.

The first question in the survey directly related to bike station locations asked the respondents to select their three preferred locations out of 12 total locations displayed on a map. These preferred locations fell into the areas with high suitability scores (Figure 10.5). In response to the first question on location, students preferred places on the ISU campus that are most frequently used, such as the Memorial Union and the State Gym (see Figure 10.5). This trend, however, was not necessarily the case for Ames residents, who reported that they preferred locations such as the North Grand Mall and the Ada Hayden Heritage Park. These results confirmed our assumption that these two groups of respondents should be treated separately.

The selection of Ada Hayden Heritage Park–the largest public park in Ames–as a desired BSP location does not coincide with high site suitability scores; it is purely based on the public participation component of the survey. This fact illustrates an advantage of combining technological models and public participation. Contrasting perceptions among different groups of people exist, proving the necessity to include both populations in the survey, and to investigate the results separately, before aggregating the final result for Ames.

To statistically test these contrasting perceptions, a T-test was conducted. A T-test is a statistical hypothesis test used to test whether two sets of data are significantly different. The objective was to determine whether the two sets of data (ISU students and Ames residents) were statistically significantly

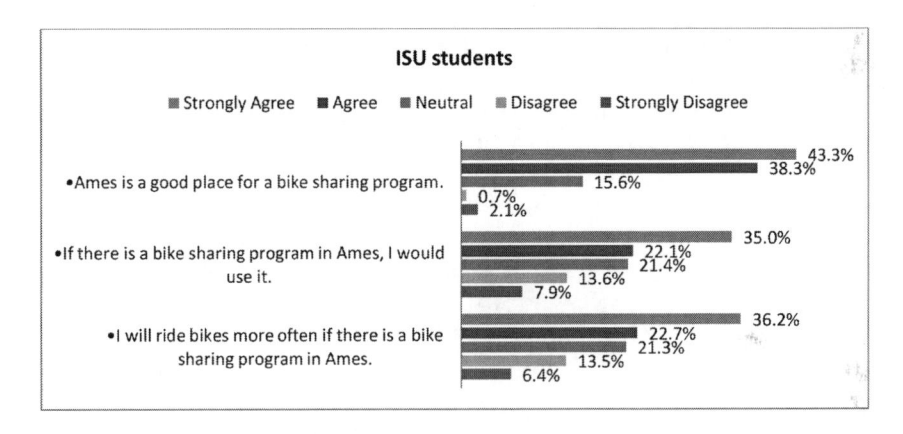

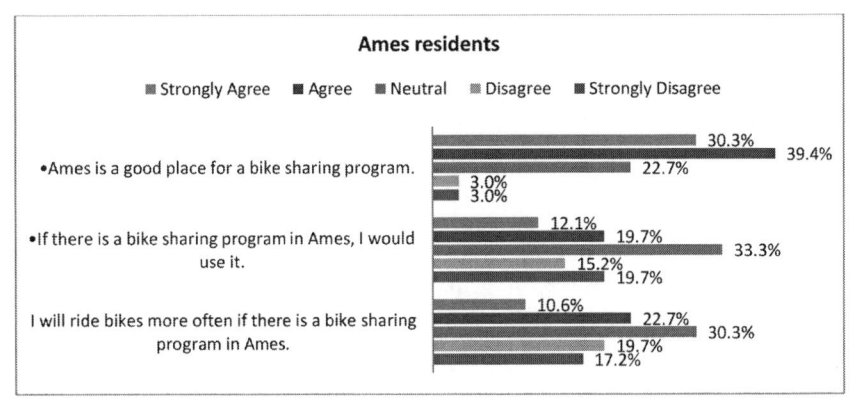

Figure 10.4a–b Level of agreement about bike-sharing programs in Ames.

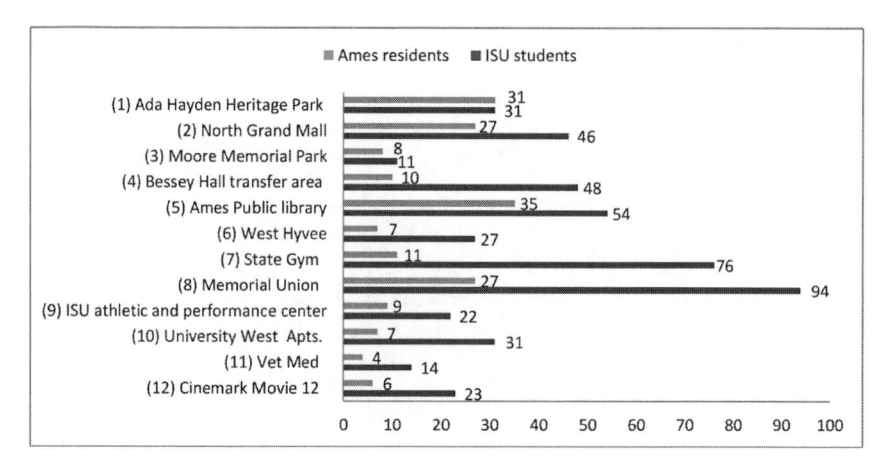

Figure 10.5 Frequency of locations chosen by respondents.

different. We conducted this test to decide whether or not to aggregate the results of preferred locations. The null hypothesis we tested was that there was no statistically significant difference between the sample of ISU students and Ames Residents. With a p-value of 0.003, the alternative hypothesis was accepted, indicating that the two data sets were statistically different and it was appropriate to report them separately.

The second survey question on bike station locations asked respondents to rank their priorities when considering placing bike stations. The possible options were: work places, attractions, neighborhood parks, transit bus stops, bike facilities, and schools. A total of 192 respondents completed a ranking. Table 10.5 displays the sum of the ranking, based on a scale from 1 to 6, where 1 was the most important and 6 was the least important. The final ranking of the six physical features is also displayed in parentheses. The smaller the sum of all data entries, the higher the rank it has been given. Public transit stops, receiving lower ranks in both groups, were determined to be the physical feature next to which bike stations should be most closely located. Interestingly, bike facilities were not perceived as a priority when compared to other choices for both groups. Even though when transportation planners develop demand models, design features are often considered as a trip generator, the public's perception of their priority may not be as high as assumed. Possible explanation could be that survey respondents identify bike facilities as accommodation already in place, thus minimizing the selection of bike facilities as a preference. It is important to highlight that schools were the highest priority for ISU students and the lowest for Ames residents.

In order to include public participation in the site selection model, five raster layers were overlaid using the weighted sum tool: work places,

Table 10.5 Sum rank of priorities when locating bikes

	ISU students	*Ames residents*	*Weight*
Work places	475 (3)	209 (5)	1.5
Special destinations	510 (4)	187 (3)	1.5
Neighborhood parks	541 (5)	166 (2)	1
Public transit stops	412 (2)	141 (1)	2
Bike facilities	560 (6)	208 (4)	0.5
Schools	400 (1)	221 (6)	NA

attractions, neighborhood parks, public transit stops, and bike facilities. Schools were not included because the data was not available at the time of this study. Each of the raster layers received a different weight, based on the final combined ranking of ISU student respondents and Ames resident respondents. For instance, the largest weight of the two was given to public transit stops because it received first ranking for Ames residents (141 respondents ranked it first) and second ranking for ISU students (412 ISU students ranked it second). Bike facilities were given a weight of 0.5 because, as a category, it ranked as the lowest priority for Ames Residents and the fourth priority for ISU students (out of 12). This question about ranking

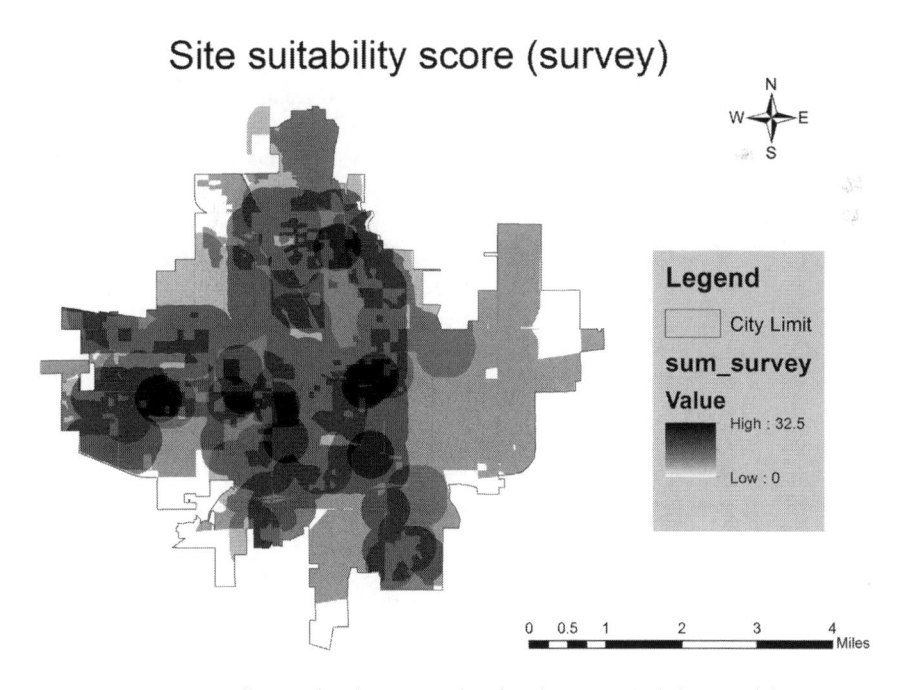

Figure 10.6 Unequally weighted sum overlay for the site suitability model.

priorities when considering placing bike stations was designed to provide justifications of how to combine GIS and public participation for this analysis. Figure 10.6 shows the overlay results of the five weighted layers based on the column "Weight" from Table 10.3. The result resembles that of the equal weighted model (see Figure 10.3).

Conclusions and limitations

This study utilized a mixed methods approach to identify potential locations for bike-share stations in the college town of Ames, Iowa, by combining GIS modeling and community surveys. Currently, for both the City of Ames and Iowa State University, a bike-share program (BSP) is only one possible strategy to promote more sustainable transportation options. Still, it is worth considering this option for its relatively low fiscal and temporal cost. BSPs can create short-term and long-term benefits to communities, such as mitigating traffic impacts, increasing transportation modal share in less fuel-consumption modes, encouraging alternative commuting means, and promoting healthy living.

Findings from the GIS model show that suitable areas where bike stations might best be located are situated along the major arterial (Lincoln Way) that connects the east and west sides of the city. Moreover, the northern part of town, where a commercial hub is located, also received a high site suitability score. The survey revealed several additional interesting patterns. When it comes to identifying preferable locations, ISU students would like destinations located within the campus. Whereas, the Ames residents would like bike stations to be more scattered throughout the city. Residents also prefer stations located close to major public attractions, such as shopping centers, parks, and the public library. One possible explanation for the difference in results might be the fact that the students see the University campus as a destination. On the other hand, residents perceive the city as a whole. For college towns looking to include bike-sharing programs in their future endeavors, the contrasting results in perception, attitudes, and preferences towards BSPs are vital to understand and to treat both populations appropriately.

Considering the unique characteristics of college towns, there are various possibilities of how a BSP in Ames could become a reality. The suitability maps (Figures 10.3 and 10.6) and survey results can help planners develop different scenarios to work with. For instance, if the University takes the lead and begins a BSP on campus, the top three locations should be Memorial Union, State Gym, and the Bessey Hall transfer area. These locations were chosen because, in addition to being located in the area of high suitability scores, they represent important landmarks within the campus. Moreover, students ranked them very highly.

On the other hand, if the city takes the lead, the most desired locations are Ames Public Library, North Grand Mall, and Ada Hayden Heritage Park. In the case of an equal partnership between the city and the university, a consensus could be reached on a plan based on the desired overall number of stations, the budget, and an implementation schedule. Finally, even though the survey results cannot be generalized to the whole population, a clear trend shows that the students favor a BSP more strongly than do residents, indicating that the University should be the leader of this process.

Some limitations need to be addressed in case other college towns are interested in using the methodology presented here to plan for bike station locations. First, the low response rates for both surveys did not represent the whole population of students or residents. The mail-in survey response rate was 17.4 percent and online survey response rate was 41.7 percent, leading to a less representative result for the mail-in survey. Second, the variables "median household income" and "alternative commuters" came from the American Community Survey and were therefore estimations. As a consequence, they were only available at the Census tract level. For a small study area such as Ames, the census tract is not ideal as the spatial unit. In addition, both variables had a margin of errors higher than 12 percent, indicating non-reliability.

In summary, the uniqueness of the study described in this chapter is the joining together of public participation with GIS modeling. It is important to understand who could be the potential users for a BSP. By separating different populations such as students and residents, planners can work under different scenarios, using appropriate tools, to reach different types of stakeholders. As a result, the final station site selection should be driven by a combination of GIS modeling results, community survey results from both students and residents, and suggestions from planners. Planners should lead this mixed methods approach because as stated previously, both GIS modeling and community surveys have their pros and cons, and need to be analyzed in relationship to one another. Finally, the GIS modeling piece of this approach can be easily replicated for other college communities, as long as socioeconomic data and shapefiles are available and reliable. Different college towns will very likely have different results from their community surveys and different weights can be captured by the spatial overlay, making this mixed methods approach flexible and effective.

Notes

1. This variable is taken from the American Community Survey. For the purpose of this study, we establish the range of values for this variable based on an assumption that people of moderate to higher incomes would (at least initially) be most interested and able to participate in a bike share program. This is not meant to convey that someone of lower income is less likely, or less deserving, to participate in bike share. It is a limitation of the study; an area that we hope will be explored in more depth in future studies.
2. For example, the scale "very good" for "population" was initially set at 997–2,355, and with the increase of 5 percent, it was set at 1,046–2,355. For "bike paths," the SA process was different. Instead of changing reclassification intervals, we changed the cell size from 10 to 15, 20, and 25 meters.

References

American Community Survey. 2010. "1-year Estimate." United States Census Bureau.

Balsas, Carlos J. L. 2003. "Sustainable Transportation Planning on College Campuses." *Transportation Policy* (10): 35–49.

Bond, Alex and Ruth Steiner. 2006. "Sustainable Campus Transportation through Transit Partnership and Transportation Demand Management: A Case Study from the University of Florida." *Berkeley Planning Journal* 125–142.

Bromley, Ray. 2006. "On and Off Campus: Colleges and Universities as Local Stakeholders." *Planning, Practice & Research* 21(1): 1–24.

City of Ames. 2010. *Ames Area Metropolitan Planning Organization 2035 Long-Range Transportation Plan.* Report prepared by HDR. Ames, IA: City of Ames.

Cochran, William G. 1977. *Sampling Techniques*, 3rd ed. New York: John Wiley & Sons.

Crosetto, Michele and Stefano Tarantola. 2010. "Uncertainty and Sensitivity Analysis: Tools for GIS-Based Model Implementation." *International Journal of Geographical Information Science* 15(5): 4215–4237.

Dagget, John and Richard Gutkowski. 2007. "University Transportation Survey: Transportation in University Communities." *Transportation Research Record* 1835: 42–49.

DeMaio, Paul and Jonathan Gifford. 2004. "Will Smart Bikes Succeed as Public Transportation in the United States?" *Journal of Public Transportation* 7: 1–17.

Detroit Bicycle Sharing. 2013. *Detroit Bicycle Sharing: Locate a Station.* www.detroitbicycleshare.com/page/locate (Link no longer available).

FHWA—U.S. Department of Transportation Federal Highway Administration. 2012. *Bike-sharing in the United States: State of the Practice and Guide to Implementation.* Report prepared by Toole Design Group and the Pedestrian and Bicycle Information Center. www.sustainablecitiesinstitute.org/Documents/SCI/Report_Guide/Guide_BikeSharingUS_Sept2012_1.pdf.

García-Palomares, Juan C., Javier Gutiérrez and Marta Latorre. 2012. "Optimizing the Location of Stations in Bike-Sharing Programs: A GIS Approach." *Applied Geography* 35(S1–2): 235–246.

ISU Live Green! 2009. *Sustainability Vision.* www.livegreen.iastate.edu/about/plan-and-vision.

Jenks, George F. 1967. "The Data Model Concept in Statistical Mapping." *International Yearbook of Cartography* 7: 186–190.

Krykewycz, Gregory, Christopher Puchalsky, Joshua Rocks, Brittany Bonnette and Frank Jaskiewicz. 2010. "Defining a Primary Market Area and Estimating Demand for a Large-Scale Bicycle Sharing Program in Philadelphia." 89th Annual Meeting of the Transportation Research Board, January 10–14, Washington, D.C.

Maurer, Lindsay K. 2012. "Feasibility Study for a Bicycle Sharing Program in Sacramento, California." 91st Annual Meeting of the Transportation Research Board, Washington, D.C.

NYC DOT—New York City Department of Transportation. 2013. *NYC Bike Share: Designed by New Yorkers.* www.nyc.gov/html/dot/downloads/pdf/bike-share-outreach-report.pdf.

Ramirez, Raul. 1996. "Spatial Data Revisions: Toward and Integrated Solution Using New Technologies." *International Archives of Photogrammetry and Remote Sensing* XXXI: 677–683. Vienna, Austria.

Rybarczyk, Greg and Changshan Wu. 2010. "Bicycle Facility Planning Using GIS and Multi-Criteria Decision Analysis." *Applied Geography* 30: 282–293.

Schäfer, Andreas, John B. Heywood, Henry D. Jacoby, and Ian A. Waitz. 2009. *Transportation in a Climate-Constrained World.* Cambridge, MA: The MIT Press.

Shaheen, Susan, Stacey Guzman, and Hua Zhang. 2010. "Bikesharing in Europe, the Americas, and Asia: Past, Present and Future." 89th Annual Meeting of the Transportation Research Board, January 10–14, Washington, D.C.

Tang, Didi. 2010. "Bike-Sharing Programs Spin Across U.S. Campuses." http://usa-today30.usatoday.com/news/education/2010-09-21-college-bike-sharing_N.htm.

Toor, Will and Spencer Havlick. 2004. *Transportation and Sustainable Campus Communities: Issues, Examples, and Solutions.* Washington, D.C.: Island Press.

TransLink—Metro Vancouver TransLink. 2008. *TransLink Public Bike System Feasibility Study.* Report prepared by Quay Communication, Inc. Vancouver BC: Metro Vancouver TransLink.

U.S. Census—U.S. Department of Commerce Census Bureau. 2010. "2010 Census Questionnaire Reference Book." www.census.gov/2010census/partners/pdf/lang files/qrb_English.pdf.

U.S. Census Viewer. "Ames, Iowa Population: Census 2010 and 2000 Interactive Map, Demographics, Statistics, Quick Facts." United States Census Bureau. http://censusviewer.com/city/IA/Ames.

11 How GPS route data collected from smartphones can benefit bicycle planning

Joel L. Meyer and Jennifer C. Duthie

Introduction

Increasingly communities are looking for ways to promote bicycling as a strategy to enhance personal mobility, reduce traffic congestion, improve environmental quality and public health, as well as for a host of other reasons (Litman et al. 2013). In order to determine the most effective ways to increase ridership in their communities, bicycle planners require data on how and where cyclists are riding on the local street network. Traditional data collection methods, including volume counts, stated preference surveys, and household travel surveys, may not provide data at the level of detail needed to fully understand local cycling behavior. Fortunately, a new data collection method has emerged in recent years that can provide planners with higher resolution route data at lower cost than has previously been possible with traditional data collection methods: the use of global positioning system (GPS) sensors in smartphones. This technology provides highly accurate traces of the routes taken by cyclists both on and off the street network, providing insight into how and where local cyclists tend to ride. This data also can be analyzed by age, gender, cycling experience or other user characteristics to give planners a better understanding of how to address the needs of different types of cyclists in the community.

Researchers at the University of Texas recently evaluated the usefulness of one such smartphone application, called CycleTracks, to collect bicycle route data. More than 3,600 unique trips were collected from around 300 cyclists in Austin, Texas, between May and October 2011 (Hudson et al. 2012). While researchers found the CycleTracks app to be useful for collecting a large dataset, to this point there has been only limited analysis of the route data in terms of its usefulness in the planning field. This chapter seeks to fill this gap by demonstrating a number of ways in which GPS route data can be used to answer specific planning questions that are difficult to analyze using current data collection methods. Specifically, it shows how GPS route data can be used to plan for network connectivity, to identify potential barriers in the bicycle network, and to analyze cycling behavior before and after the installation of a new facility. All of the procedures presented in this chapter can be implemented using geographic information system (GIS) software available to most local planning agencies. While Austin, Texas, is used as a case study, it is the authors' hope that other cities can adopt the methods presented in this chapter in order to make better informed planning and policy decisions regarding bicycling in their communities.

Background

Traditionally bicycle planners have relied on volume counts, stated preference surveys, facility inventories, and household travel surveys to collect information on local bicycling behavior (Pedestrian and Bicycle Information Center 2005). While these methods can provide valuable information to planners, they are limited by their inability to provide detailed data on cyclists' revealed route choices, along with issues related to sampling bias and poor respondent trip recall (Pratt, Evans, and Levinson 2012). These limitations, combined with the drastic reduction in cost of commercially available devices equipped with GPS functionality over the past decade, have led many in the transportation planning field to believe that utilizing smartphones for data collection can provide valuable new information to supplement local planning agencies' regular data collection efforts (Federal Highway Administration 2011).

In recent years, planning entities and researchers have begun to implement smartphone-based data collection efforts. Portland Metro, the metropolitan planning organization for Portland, OR, deployed an application called Pacelogger as an optional add-on to its 2011 travel diary survey (Rodriguez 2012). Researchers at the Singapore-MIT Alliance for Research and Technology developed the Future Mobility Survey, a smartphone-based travel survey application for use in the Republic of Singapore's Household Interview Travel Survey (HITS) in 2012 (Cottrill et al. 2013). More recently, researchers at the University of Minnesota measured the impacts of daily travel routines on physical activity and psychological well-being through the creation of an Android-based application called UbiActive (Fan et al. 2012).

The number of studies using smartphones to specifically study bicycling behavior, however, is very few. To the authors' knowledge, the first smartphone app developed specifically for this purpose is CycleTracks, which was developed by the San Francisco County Transportation Authority (SFCTA) in 2010 with the goal of incorporating bicycling into SF-CHAMP, their activity-based travel demand model. The CycleTracks app, which is freely available on both Android and iOS (iPhone) platforms, collects users' route patterns, distances, travel times, and trip purposes, as well as other optional information including age, gender, home and work zip code, and cycling frequency. In Phase 1 of data collection, November 2009 through April 2010, a total of 3,034 valid trips from 366 users were collected by SFCTA after cleaning and smoothing the data. Using this data, SFCTA developed a bicycle route choice model that showed that San Francisco cyclists strongly preferred bicycle lanes to other facilities, and tended to avoid steep hills, turns, and excessive deviations from their shortest path (Charlton et al. 2011). The open source app has been used and adapted for several purposes in at least 12 other cities including studying travel on the Texas A&M University campus, studying bicycling in the City of Atlanta, and for our purposes to study bicycling in the City of Austin.

As this section shows, travel data collected from smartphones has been used for a wide variety of purposes. While these are important contributions to transportation research, little work has been done in terms of showing how this data can be used to aid planners in making better informed decisions regarding bicycle planning. This chapter adds to previous studies involving GPS route data by presenting innovative ways by which the data can be used to plan for network connectivity, to identify potential barriers in the bicycle network, and to analyze cycling behavior before and after the installation of new facilities. The following section will discuss the data collected for this study, and then the final section will provide step-by-step procedures for how the GPS data can be used to perform these analyses and discuss their significance in the broader context of bicycle planning.

Data collection

Beginning in May 2011, researchers at the Texas A&M Transportation Institute and The University of Texas at Austin's Center for Transportation Research marketed the CycleTracks app to the Austin bicycling community. Over the six-month data collection period, the team was able to collect more than 3,600 trips from 317 participants. Of these users, 70 percent were male and 30 percent were female. More than one-third of users were between the ages of 20 and 29 (37 percent), 28 percent were between 30 and 39, and 22 percent were between 40 and 49. In terms of cycling frequency, the majority of users rode several times per week (44 percent), while another 39 percent rode daily. The numbers suggest that the Austin dataset is skewed towards more experienced cyclists, which was also the case in the San Francisco study. Data from the initial analysis showed that experienced cyclists were more likely to ride on streets with higher speed limits than inexperienced cyclists, that more than 60 percent of all route miles were taken on roads adjacent to single family uses, and that females were more likely than males to use local neighborhood streets. Researchers concluded that despite the significant time required to process the data after collection, the CycleTracks app was useful for collecting a large dataset at relatively low cost (Hudson et al. 2012).

Planning applications of GPS route data

Using Austin as a case study, the following sections demonstrate how GPS route data can be used to aid planners in making better-informed decisions regarding bicycle planning. Specifically, it will be shown how the data can be used to plan for network connectivity, to identify barriers in the bicycle network, and to analyze cycling behavior before and after the installation of new facilities.

Planning for network connectivity

A key component of creating a bicycle-friendly community is ensuring that all levels of cyclists can reach destinations safely and conveniently. This requires building a well-connected network that provides a range of different facilities for cyclists to use. While many cyclists prefer on-street, or "integrated" facilities due to their better connectivity and faster riding speeds, many others prefer "segregated" facilities that separate them from traffic entirely. The diverse range of preferences on the part of cyclists means that planners should strive to make the entire street network safe for cycling while also developing key routes that are particularly fit for cycling, either due to low traffic volumes, fewer hills, or physical separation from traffic (Litman et al. 2013). Without adequate data, however, determining which streets are key for connectivity is a difficult task. A benefit of GPS data, then, is that it reveals which routes cyclists use most often, as well as the most direct connections between their origins and destinations. The method presented in this section demonstrates how these two pieces of information can be used to aid planners in identifying key bicycle corridors to create a complete bicycle network in the city.

Creating the shortest path and volume maps

This method uses a relatively straightforward procedure in ArcGIS 10.1 to identify key bicycle routes in Austin, and involves only three datasets. The first is the street network, which was created from the City of Austin's publicly available street-centerline shapefile (a geospatial vector data format for geographical information system software) with modifications made to represent facilities likely to be used by cyclists, such as off-street paths, sidewalks, and parking lots. In all, 923 additional links were manually digitized to the street shapefile to better model cycling behavior (see Hudson et al. 2012 for a detailed description of the digitizing process). The only other data needed for this method are the GPS traces and their corresponding shortest paths. Shortest paths were created by finding the shortest network distance between each trip's origin and destination using the ArcGIS Network Analyst extension, with automatic iterations performed for each of the 3,090 matched trips using a custom-built tool in ModelBuilder, a programming tool that allows the user to create new functionalities within ArcGIS.

The first step in this procedure involves spatially joining the GPS traces and shortest paths to the Austin street network to determine the number of times they traversed each street segment over the study period. This results in two network shapefiles that can be used to evaluate network connectivity. Symbolizing the newly created volumes and shortest path attributes provides us with a visualization of the most heavily used streets over the study period and the street segments that provide the greatest connectivity for users, respectively. The Volume Map in Figure 11.1 and the Shortest Path Map in Figure 11.2 show the observed number of trips that traversed each street segment and the number of trips whose shortest path occurred on each street segment, respectively, over the study period.

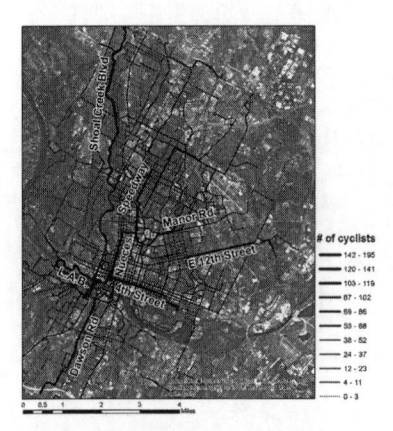

Figure 11.1 Volume map.

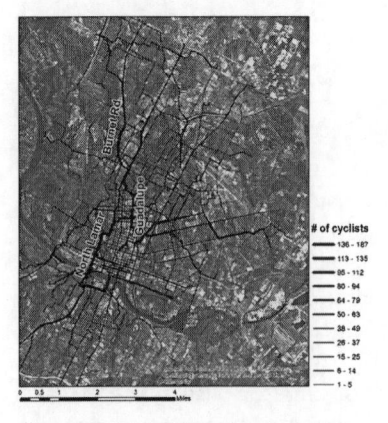

Figure 11.2 Shortest path map.

Discussion of the Shortest Path and Volume Maps

Looking first at the Volume Map (Figure 11.1), we can see that three of the most heavily used routes (based on the dark gray coloring) are the Lance Armstrong Bikeway/4th Street, Shoal Creek Boulevard, and Nueces Street. These routes provide interesting insight into bicycling behavior and network planning in Austin. With a $3.2 million price tag and named after one of the city's most famous athletes, the Lance Armstrong Bikeway (LAB) is arguably the most high-profile bicycle facility in Austin. When completed, it will provide a much needed, six-mile connection between west and east Austin, with the majority of its length on separated bike paths. Interestingly, the Shortest Path Map (Figure 11.2) does not show the same dark coloring along this stretch as was seen in the Volume Map. This indicates that cyclists are using the LAB despite the fact that it is not necessarily their most direct connection. In fact, of the 178 trips that traversed the LAB from Trinity Street to Interstate 35 over the study period, only five (2.8 percent) had their shortest paths along that same stretch. These trends seem to indicate that investing in separated facilities provides an attractive option for cyclists, which confirms what we know from stated preference surveys. From a planning standpoint, the Volume and Shortest Path maps can provide Austin planners and politicians with a powerful visual aid to help justify public spending on bike infrastructure and to garner support for similar projects in the future.

Another heavily used route seen in the Volume Map is Shoal Creek Boulevard (Figure 11.3). This north–south corridor appears in black or dark grey for over four miles in north central Austin. If we compare the Volume Map to the Shortest Path Map, however, we can see that overall, cyclists' most direct north–south connection was on Burnet Road/Medical Parkway. Indeed, of the 93 trips made along Shoal Creek Boulevard from Allandale Road to 38th Street, 66 had their shortest paths on Burnet Road/Medical Parkway. By taking Shoal Creek Boulevard rather than their shortest paths along this stretch, cyclists added an average of 0.35 miles to their total trip distance (i.e., increasing their trip length 14.5 percent). A quick comparison of these two routes in Google StreetView suggests that Shoal Creek Boulevard has wider bicycle lanes, lower traffic volumes and speed limits than Burnet Road/Medical Parkway. From a planning standpoint, the fact that bicyclists are not using their shortest path along this stretch is not a problem per se, but if we believe trip length to be an important factor for whether or not people choose to bike, then planners might want to explore ways in which the Burnet Road/Medical Parkway corridor can be improved to provide increased connectivity through this area.

The examples presented in this section have shown how GPS route data can aid planners in identifying key cycling routes in the city. The Volume and Shortest Path maps can be especially useful to planners for use in community stakeholder meetings to visualize high-level bicycling trends.

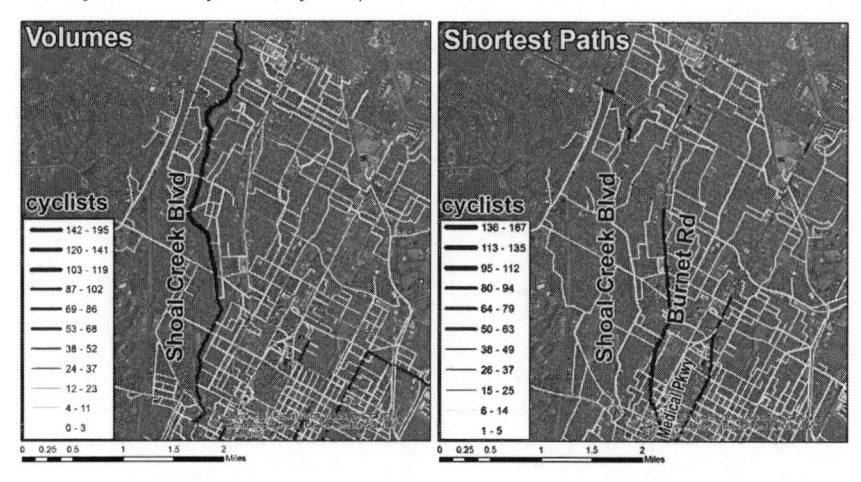

Figure 11.3 Shoal Creek Boulevard volumes (left) and shortest paths (right).

The Volume Map, for example, can act as a visual aid for planners to use when skeptics call into question the value of investing in bike infrastructure. These maps can also be used for sketch planning exercises whereby stakeholders can visualize current conditions and interactively mark their own ideas on network connectivity on the maps. Whatever the application may be, the high-level visualization of local ridership patterns that GPS data provides is a key benefit over traditional data collection methods.

Barrier identification

Another key activity involved in bicycle planning is identifying and prioritizing specific locations in the bike network in need of planning or policy attention. Barriers to bicycling can result from a lack of facilities for bicyclists, poor connectivity in key locations, subpar pavement conditions, and a host of other real and perceived obstacles. Because planners cannot be expected to be aware of all barriers in the network at all times, the Pedestrian and Bicycle Information Center recommends that, "a systematic procedure is needed to identify what (and where) countermeasures should be implemented" (Pedestrian and Bicycle Information Center 2013). In addition to identifying potential projects, planners should develop a prioritization system to rank competing projects given limited funding for bicycle infrastructure. To accomplish these tasks, planners would benefit from better information on where barriers to bicycling might be occurring in the local bike network. This section will present two methods by which GPS route data can be used to identify potential barriers in the local bike network. The first method identifies barriers at the network level, while the second method identifies barriers at the block level.

Creating network-level barrier maps

Both procedures presented in this section identify barriers in the Austin bicycle network by finding locations where the GPS traces show repeated deviations by bicyclists from the shortest possible path they could have taken. The assumption is that deviations from the shortest path indicate that there is something about that particular location that deters cyclists from traversing it, and that planning or policy intervention might be necessary. Conversely, deviations from the shortest path at a particular location might signal that cyclists are willing to go slightly out of their way to ride on other nearby facilities, and that current infrastructure is actually adequate. In either case, these methods provide a crucial starting point for systematically assessing potential barriers to bicycling. While deviations for the shortest path by an individual cyclist may not say much, with more than 3,000 trips in the CycleTracks dataset, repeated deviations from the shortest path reveal "hot spots" in the bike network that warrant closer examination.

The first method uses the Austin street network as the unit of analysis for identifying barriers. It borrows from the Bicycle Network Planning method presented in the previous section, but goes one step further: Rather than looking at an aggregate view of the most heavily used routes and shortest paths, this method looks at the difference between these two values for each street segment in the study area to determine potential barriers to bicycling.

First, we merge the matched routes and shortest paths into one feature class, which will allow us to determine the difference between the number of times bicyclists used each street segment ("matched trips") and the number of times their shortest paths traversed each street segment ("shortest paths"). Using the Add Field tool, we can create a new attribute field called "Difference," and use the Field Calculator tool to fill this column with the following formula: Difference = (matched trips—shortest paths). The resulting value indicates whether cyclists tended to take or avoid their shortest paths for each street segment in the study area. Figure 11.4 shows the resulting map in Central Austin, with the "Difference" field symbolized. Darker and thicker lines indicate street segments with a large difference between the number of bicyclists using it and the number of times it appeared on a shortest path.

Network-level barrier maps

By looking at the street segments with the highest level of avoidance, we begin to get an idea of where potential barriers might exist in the Austin bike network. The top four street segments with the largest differentials all occur along West Avenue in central Austin. Of the 135 cyclists whose shortest paths occurred along this entire stretch, only one took it. Looking closer at these streets in Google Street View, it is not immediately apparent why so many cyclists chose to avoid this area; while there is a slight hill along this stretch, these streets go through a residential neighborhood with relatively low traffic. If we zoom out and look at the streets adjacent to West

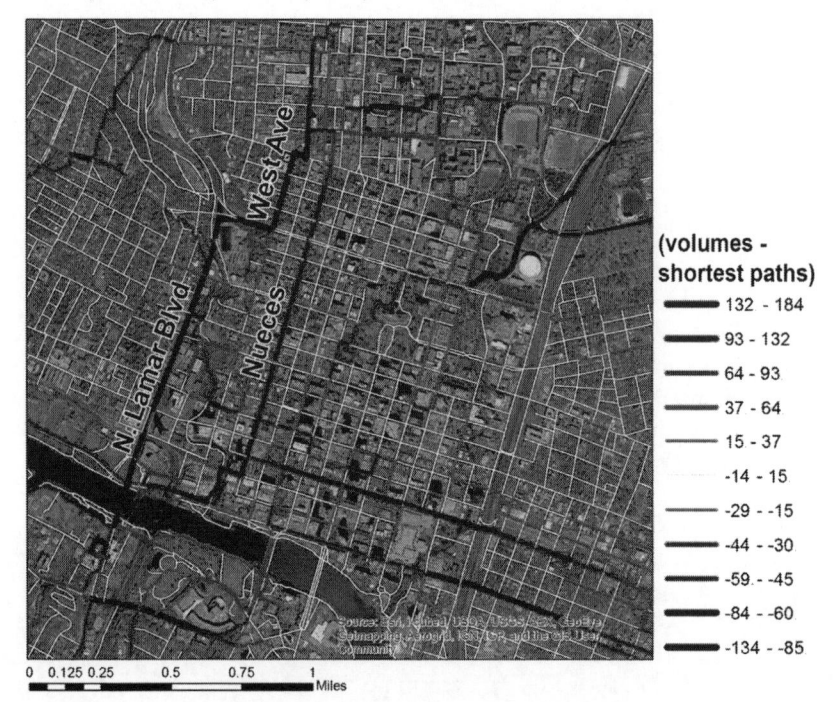

Figure 11.4 Network differences map, central Austin.

Avenue, however, we can see that Nueces Street is located only two blocks to the east, which we already know from the previous section to be an attractive north–south route for bicyclists. Indeed, of the 135 cyclists whose shortest paths occurred along West Avenue, 49 of them took Nueces Street instead. Based on this information, it seems as if the high level of avoidance seen on West Avenue had nothing to do with the physical attributes of the street itself, but rather the relative attractiveness of Nueces. In other words, we can probably rule out West Avenue as a significant barrier to bicycling.

The West Avenue example showed that although cyclists are avoiding certain street segments, there are alternate routes available to them in these areas (i.e., Nueces Street). This pattern of highly avoided streets that are nearby streets with high actual ridership can be seen throughout the study area. The arrows in Figure 11.5 show areas where bicyclists show preference for one route (darker color) over their shortest route (lighter color).

It is not enough to simply look at the difference field to identify potential barriers. Instead, we must look for highly avoided areas that do not have clear alternate routes nearby. One such area that can be seen from the Difference Map is Pleasant Valley Road in East Austin, from Lyons Road to East 12th Street (Figure 11.6). Of the 66 cyclists whose shortest paths traversed this stretch, only nine took it.

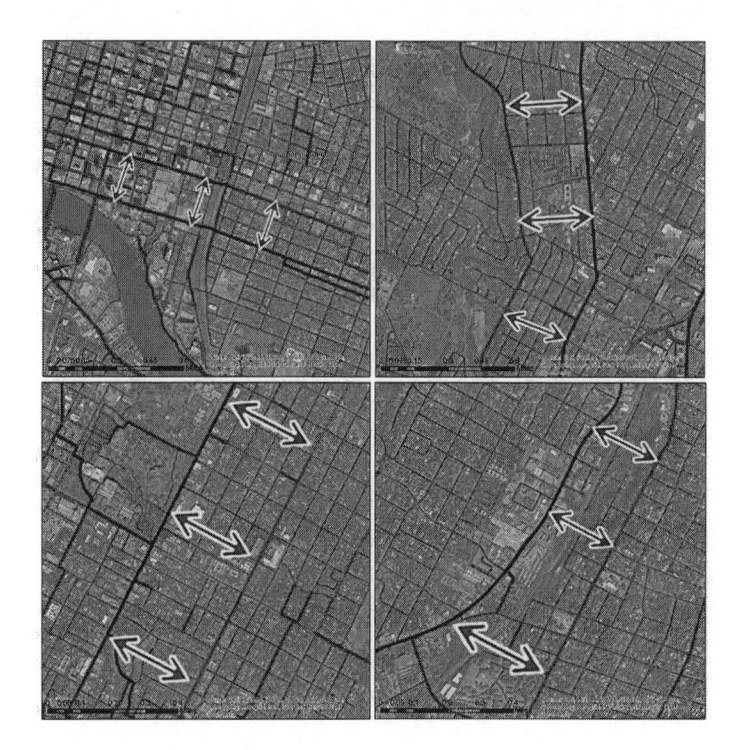

Figure 11.5 Volume and shortest path pairs.

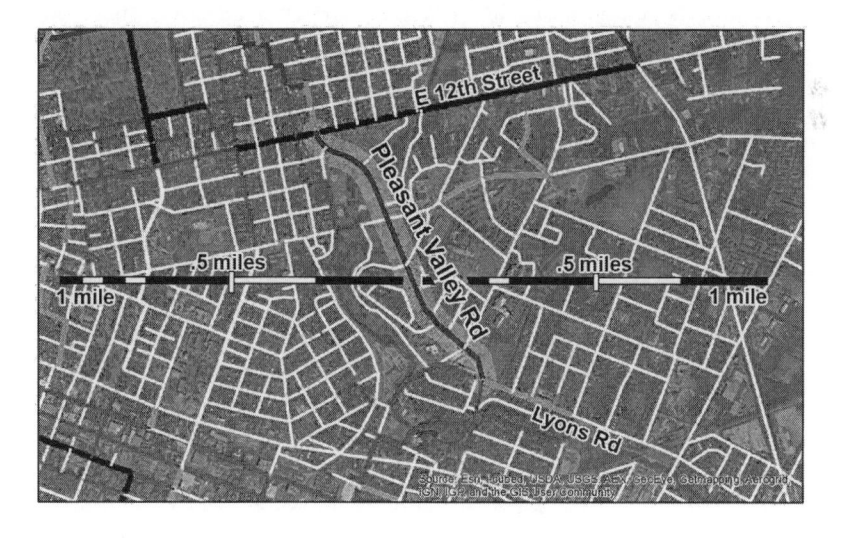

Figure 11.6 Pleasant Valley Road.

From Figure 11.6 it is not immediately apparent which alternative routes cyclists were taking when avoiding Pleasant Valley Road. In other words, in contrast to the examples of avoidance that were caused by attractive alternative routes, it appears that the high level of avoidance of Pleasant Valley Road has something to do with the street itself. This signals that it is a potential barrier to bicycling. The next question is to what degree is Pleasant Valley affecting rider behavior. To answer this, we can sum the trip lengths of actual routes taken and compare them to the shortest path trip lengths. The 66 trips that avoided Pleasant Valley Road had an average length of 4.47 miles, compared with their corresponding shortest paths, which had an average length of 3.70 miles. This means that cyclists who avoided Pleasant Valley went an average of 0.77 miles out of their way over the course of their trip (an increase of 20 percent). The repeated avoidance of Pleasant Valley Road and subsequent longer trip lengths means that the road should be looked at further to determine whether or not it is a barrier to cycling in this area. It is likely that the long hill along this stretch, high speeds and traffic volumes, and lack of bicycle facilities are all contributing to the high level of avoidance on Pleasant Valley Road.

Creating block-level barrier maps

In contrast to the network barrier identification method, which identified specific street segments that cyclists avoided, the block-level barrier identification method identifies specific areas avoided by bicyclists. This method uses the same assumption as the first—that repeated deviations by cyclists from their shortest paths represent potential barriers to cycling—but provides slightly different results that can be used by planners to supplement their barrier identification.

Figure 11.7 illustrates the basic procedures involved in the block-level barrier identification method. The first step involves joining each trip and its corresponding shortest path into single features (Steps 1 and 2 in Figure 11.7). Once each trace/shortest path pair is merged together, we can create polygons in between the gaps formed by each pair of lines (Step 3 and Step 4). For a given trip, these gaps represent areas where a cyclist's route deviated from their shortest path. To find areas cyclists frequently avoid, we can simply overlay each of the trace/shortest path gaps (polygons) on top of one another (Step 5). We can then use the Dissolve tool to count how many times these polygons overlap one another (Step 6).

After these six steps have been conducted, we are left with a dataset that includes polygons with attribute values indicating the number of times cyclists avoided their shortest path in that particular area. Figure 11.8 shows the resulting dataset, with higher attribute values (i.e., areas avoided by bicyclists) symbolized in darker colors.

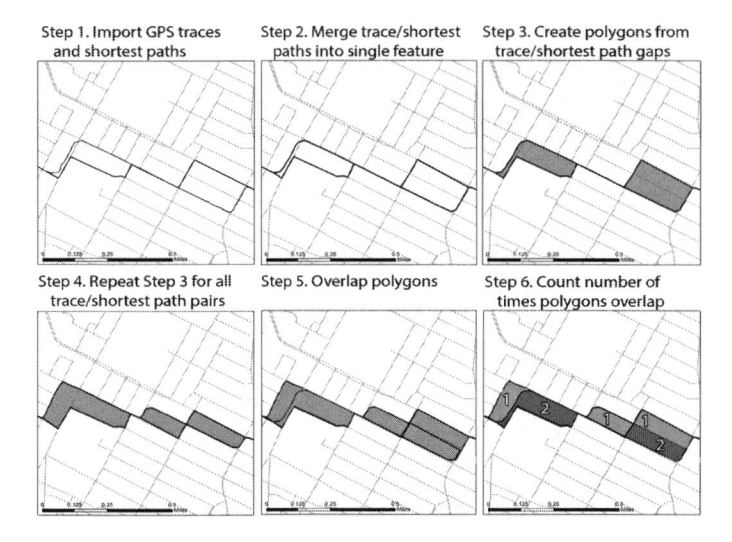

Figure 11.7 Block-level barrier identification procedure.

Discussion of the block-level barrier map

For this method, it is tempting to conclude that the areas with the highest attribute values (i.e., the highest number of overlaps) indicate the most significant barriers in the bike network, but this can be misleading. Because more overlaps inevitably occur in areas where cyclists ride the most overall, looking only at the aggregate number of overlaps will lead to the conclusion that significant barriers only occur in high-traffic areas. To account for this bias we must look at smaller geographic areas to identify more localized barriers to bicycling.

To provide an example of how this method can aid planners in identifying barriers to bicycling, The University of Texas at Austin campus is used as a case study. We first create a half-mile buffer around the boundaries of the UT campus to isolate the study area. Re-symbolizing the attribute values to include only the values in the study area reveals barrier hot spots in more detail (Figure 11.9, top left). Zooming in closer on one of the hot spots (top right), the data suggests that there is a potential barrier occurring between Inner Campus Drive and 21st Street. Further analysis can be performed to determine why a hot spot is occurring between these two streets. Activating the shortest path and GPS trace layers reveals an interesting trend in this area: Of the 69 routes whose shortest path went through Inner Campus Drive, zero cyclists chose to use this route (Figure 11.9, bottom left).

The fact that zero cyclists used Inner Campus Drive during the study period despite the fact that it was their shortest path indicates a potential barrier to bicycling on this section of the UT Campus. There are a number

High ▬▬▬▬▬▬▬▬▬▬▬▬▬▬▬▬▬▬▬▬▬▬▬ Low
Level of avoidance of cyclists' shortest paths

Figure 11.8 Barrier map.

of reasons why cyclists might avoid Inner Campus Drive. For one, the road requires cyclists to climb a steep hill when entering from Speedway to the east, which may deter many potential cyclists from accessing the road. Another potential reason why cyclists might be avoiding Inner Campus Drive is that they are required to ride into oncoming vehicular traffic when riding east and north. While the road allows bicyclists to travel in both directions, it is designed for one-way car traffic with parking along one side (Figure 11.9, bottom right). This configuration is likely uncomfortable for less experienced cyclists. From a planning standpoint, it is possible that minor changes to the street configuration, eliminating on-street parking, or providing better signage would allow cyclists to feel more comfortable using

Figure 11.9 Barrier analysis, UT Campus.

this road. Obviously distance is only one factor that goes into cyclists' decision to ride or not, but the fact that all 69 cyclists chose to avoid their shortest path is a telling sign that Inner Campus Drive might be a barrier to increased bicycling on the UT campus.

Before and after analysis

The ability to understand how new facilities affect rider behavior is key to effective bicycle planning, especially in a field where performance metrics play an important role in securing funding for projects. As was discussed earlier, however, traditional data collection methods may not be capable of providing data at the level of detail necessary to fully evaluate the effectiveness of new facilities in promoting cycling. Before/after studies involving user opinions or volume counts, for example, can only provide a general idea of the value of new facilities. A clear benefit of GPS data, then, is that it reveals more nuanced behaviors, allowing planners to show how new facilities actually benefit cyclists. This section will use a high-profile bicycle facility in Austin—the South Congress Improvement Project—to demonstrate how GPS route data can be used to analyze cycling patterns before and after the installation of a new cycling facility.

South Congress Avenue is an iconic street in Austin, home to an eclectic mix of restaurants, music venues, and boutique retail stores. To address problems associated with increased congestion, a lack of parking, and poor bicycle and pedestrian facilities, Austin voters passed a $4 million bond referendum in 1998 to fund the South Congress Improvement Project. After a public input process and many design concepts, the final project included sidewalk and curb ramp repairs to meet ADA standards, the retiming of traffic signals to reduce travel speeds, the installation of bulb-outs at intersections to shorten crossing distances for pedestrians, a consolidation in the number of bus stops from 12 to six to improve transit service, and the implementation of back-in angled parking spaces to increase the number of parking spaces and reduce the risk of collisions. Bicycle-specific improvements included the installation of bike lanes along the uphill (southbound) and flat sections, and shared-use lanes with sharrows along the downhill (northbound) sections. Additional bicycle parking was also installed (Schatz 2012).

GIS procedure for before and after analysis

Because we are interested in analyzing cycling behavior before and after a known date, the first step is to separate the trips into two subsets based on when they occurred. Using the Select By Attributes tool we can simply query those trips made before and after July 7, 2011 (the date when the South Congress improvements were completed), and create new shapefiles for each set. Performing this operation reveals that of the 3,090 trips made during the study period, 1,503 (48.6 percent) occurred before the South Congress Avenue improvements and 1,587 (51.4 percent) occurred after. This relative balance in the number of before and after trips is to be expected as the improvements were completed right in the middle of the study period (May 1 to October 31, 2011).

Before and after analysis results

Once we have divided the trips by their date, we can look more closely at each subset to see whether there are differences in riding behavior before and after the improvements were made. A simple test is to count the number of before/after trips made along all or part of South Congress in order to get a general idea of whether or not the improvements led to increased ridership. Using the Select by Location tool, we can select those trips that share a line segment with the 1.33-mile stretch of South Congress where new facilities were installed. This operation reveals that of those trips that traversed at least one street segment along the South Congress study area,

24 occurred before the new facilities and 36 occurred after. This represents a 50 percent increase in ridership. Because we know that overall study period ridership was about equal before and after July 7, this finding seems to indicate that there was indeed a substantial increase in ridership after the improvements were made to South Congress.

Another way to analyze before and after ridership patterns is to employ the network-based Barrier Identification procedure presented in the previous section. Using this procedure will allow us to investigate how the new facilities might have influenced cyclists' decisions to choose or avoid their shortest paths. Instead of looking at the trace/shortest path differences over the entire study period, however, this analysis requires that we first create separate datasets for before and after trips. Once this is done, the traces and shortest paths can be spatially joined to the street network using the same procedure described in the Network Planning method. Figure 11.10 shows the trace/shortest path differentials for each of the 18 street segments along the South Congress Avenue study area before and after the improvements were made.

A noticeable difference between the before and after route data occurs between Gibson Street and E. Riverside Drive (Figure 11.11). Before the improvements, the data shows that a number of cyclists avoided taking this stretch (Figure 11.11, top left), despite the fact that it was their shortest path. After the improvements were made, however, this disparity disappears entirely (Figure 11.11, bottom left). Looking at the before and after views in Figure 11.11, we can see that the South Congress improvements included the installation of a climbing lane along the southbound (uphill) side of this segment, and the installation of bike lanes along the northbound (downhill)

Figure 11.10 South Congress barriers before and after improvements.

Figure 11.11 Riding behavior after installation of climbing lane (triangles point to location where photographs were taken).

section. While it is not possible to determine causality with the small sample available, there does appear to be some correlation between the South Congress Avenue improvements and bicyclists' decision to ride on their shortest paths along this stretch.

The methods presented in this section can benefit local bicycle planners in a number of ways. The ability to understand how new facilities affect riding behavior is key to effective planning in terms of getting the most out of limited resources. In addition, if planners can provide tangible evidence that investing in bicycle facilities leads to a better experience for cyclists then they are more likely to garner public and political support for projects in the future.

Conclusions

The previous section demonstrated three ways in which GPS route data can be used to aid planners in the bicycle planning process. Specifically, we show how the data can be used to identify key bicycling corridors to plan for a well-connected network, how to identify potential barriers to bicycling, and how to analyze cycling behavior before and after the installation of new facilities to help determine the effectiveness of bicycle infrastructure investments. These methods showed how the spatial and temporal information provided by GPS route data can give planners a more detailed understanding of bicycling behavior than has previously been possible with traditional data collection methods. As communities look for ways to promote bicycling, this information can allow them to make more informed decisions regarding infrastructure investments and policies that will help increase bicycling comfort and safety.

While smartphone-based data collection methods can benefit planners in ways that traditional data collection methods cannot, they possess a number of limitations that should be addressed before they are adopted as a regular part of planning agencies' ongoing data collection efforts. One way in which these methods can be improved upon is to find ways to make the data collection passive rather than active. The fact that users must interact with the app each time they want to record a trip likely leads to lower participation rates and fewer trips recorded than would be if the app was running at all times. Minimizing user burden would also allow researchers and planners to gather a representative sample more effectively than with an active application. Ideally, the application would be able to automatically detect a user's trip, mode, and even trip purpose while running in the background of their phones. These features would likely lead to a quicker drain on the phone's battery life, which has also been shown to be a major limitation of these methods.

Next, researchers and practitioners should explore ways to reduce, or at least control for, sampling bias. The fact that not everyone owns a smartphone excludes certain segments of the population (namely minority and low-income groups) and is an obvious limitation of these methods. Participants in these studies have also tended to be more experienced cyclists than the cycling population as a whole. This bias was seen in the Austin CycleTracks dataset, with 39 percent of users riding daily and another 44 percent riding several times per week. The fact that experienced cyclists are participating in these studies more often than novice riders has important implications for the future of smartphone data collection and its usefulness for bicycle planning. While the "strong and fearless" will ride nearly anywhere, a key focus of bicycle planning is to determine ways to increase ridership for the "interested but concerned" demographic (Geller 2006). Future data collection efforts should find ways to recruit a larger user base of less experienced cyclists in order to understand these behavioral differences.

Finally, the time and expertise required to post-process and analyze the GPS data may be impractical for local planning agencies that do not have extra staff time or resources. In order for these methods to be adopted on a large scale, the data needs to be able to be processed and analyzed in software programs that are familiar to planners. Researchers and practitioners should therefore explore ways to create automated map-matching algorithms and other open-source data cleaning tools in order to make data processing simpler for local planning agencies.

References

Charlton, Billy, Elizabeth Sall, Michael A. Schwartz, and Jeff Hood. 2011. "Bicycle Route Choice Data Collection Using GPS-Enabled Smartphones." Transportation Research Board 90th Annual Meeting Compendium of Papers. Washington, D.C.

Cottrill, Caitlin D., Ines F. Dias, Hock B. Lim, Moshe Ben-Akiva, and P. Christopher Zegras. 2013. "The Future Mobility Survey: Experiences in Developing a Smartphone-Based Travel Survey in Singapore." Transportation Research Board 92nd Annual Meeting Compendium of Papers. Washington, D.C.

Fan, Yingling, Qian Chen, Chen-Fu Liao, and Frank Douma. 2012. *Smartphone-Based Travel Experience Sampling and Behavior Intervention Among Young Adults*. Intelligent Transportation Systems Institute Report No. CTS 12–11. Minneapolis, MN: University of Minnesota ITS Institute.

Federal Highway Administration (FHWA). 2011. "Pedestrian and Bicycle Data Collection: Final Report." Report prepared by AMEC E&I, Inc. and Sprinkle Consulting, Inc. Washington, D.C.: FHWA.

Geller, Roger. 2006. *Four Types of Cyclists*. Portland, OR: City of Portland Office of Transportation.

Hudson, Joan G., Jennifer C. Duthie, Yatinkumar K. Rathod, Katie A. Larsen, and Joel L. Meyer. 2012. *Using Smartphones to Collect Bicycle Travel Data in Texas*. University Transportation Center for Mobility Report 11–35–69. College Station, Texas: Texas Transportation Institute.

Litman, Todd, Robin Blair, Bill Demopoulos, Nils Eddy, Anne Fritzel, Danelle Laidlaw, Heath Maddox, and Katherine Forster. 2013. *Pedestrian and Bicycle Planning: A Guide to Best Practices*. Victoria, BC: Victoria Transport Policy Institute.

Pedestrian and Bicycle Information Center. 2005. *Identify and Prioritize Locations Needing Improvement*. bicyclinginfo.org. (Now pedbikinfo.org; link no longer active).

Pedestrian and Bicycle Information Center. 2013. *Performance and Analysis*. www.pedbikeinfo.org/planning/analysis.cfm.

Pratt, Richard H., John E. Evans, and Herbert S. Levinson. 2012. *Pedestrian and Bicycle Facilities: Traveler Response to Transportation System Changes*. Transit Cooperative Research Program Report 95. Washington, D.C.: National Academies of Sciences, Engineering, and Medicine.

Rodriguez, Sandra. 2012. *Smart Phone Apps with GPS—"Pacelogger" and "RouteScout."* PTV NuStats. www.nustats.com/ January 25.

Schatz, Gary W. 2012. *Transforming South Congress in Austin, Texas*. Pasadena, CA: Institute of Transportation Engineers Technical Conference and Exhibit.

12 Mapping GPS data and assessing mapping accuracy

Katie A. Kam, Joel L. Meyer,
Jennifer C. Duthie, and
Hamza Khan

Introduction

Global positioning system (GPS)-enabled smartphones make collecting data on the routes cyclists choose easier than ever. Analyzing the large quantities of data output by such devices (e.g., millions of GPS points giving the location of cyclists every few seconds) poses a challenge, though, because of some of the "noise" in GPS data and the ability of cyclists to deviate from the roadway network. The post-data collection process requires cleaning the data (e.g., removing anomalous data, verifying bike mode, and splitting multiple "chained" trips into single trips), adding links (e.g., roads, trails, alleys, and paths through parking lots) as needed to an existing roadway network file, and then relating the remaining GPS points with the network links to identify the path most likely taken by the cyclist on the network (in a step referred to herein as "network mapping," also called map-matching in other references).

While all of those steps are critical, this chapter focuses on the last step by describing methods to map the network and check for accuracy of the network mapping effort without the use of supplemental information such as travel diaries or interviews. To clarify, the network mapping and accuracy assessment described in this chapter focuses on post-data collection analysis and not real time analysis (although the methodology may be transferable with modifications). This line of research completely changes the way transportation planning agencies collect and analyze data. Accurate network mapping of GPS data provides output useful for identifying the facilities cyclists choose to use and areas to ride through and the inputs required to develop bicycle route choice models.

Background

Because determining how a line of GPS points (i.e., a GPS trace) for each trip relates to the network links along with finding the most likely path taken by the cyclist is not an easy task, research in this area has explored various methods to use GPS points to select network links, with each method having pros and cons. Schuessler and Axhausen (2009) wrote a thorough review of the literature in this area, in which the authors categorize network mapping (map-matching) methods as geometric, topological, and advanced. Zhou and Golledge (n.d.) also provide a review of those different approaches to mapping travel routes, as well as Quddus, Ochieng, and Noland (2007), but with particular emphasis on real-time applications. The remainder of this section compares and contrasts those different categories of network mapping methods for purposes of putting into context the network mapping strategy used for the study presented in this chapter.

Geometric approaches map GPS points to the nearest point, link, or node (e.g., intersection of links or terminus of link) in a network. The simplest example of a geometric approach consists of determining the path taken on the network by selecting the network link closest to each GPS point; however, the resulting routes typically contain errors. For instance, for a GPS trace following a roadway with intersections, as the line of GPS points approaches an intersecting roadway, the geometric method would have the GPS points closest to the intersecting roadway select that roadway instead of the roadway the GPS trace followed. Use of a geometric method risks the selection of network links that are not part of the actual path taken. In addition, selecting network links closest to the GPS points could result in the formation of infeasible routes (e.g., because of one-way restrictions and dead-ends). The geometric approach cannot guarantee that the selected links create a feasible path connecting the origin and destination of the trip. In a study of car trips tracked by GPS, the use of this type of approach resulted in 33 percent of the 3,000 trips being incorrectly matched to the 300,000 links that represent the roadway network (Nielsen, Wurtz, and Jorgensen 2004). Newsom and Krumm (2009) also describe this problem. For both studies, supplemental information provided the means for checking the accuracy of the paths created by the geometric mapping method.

Geometric methods that take into account the direction of the sequence of GPS points select network links that closely follow the directional movement of the GPS trace (e.g., an algorithm mapping a GPS trace heading southeast would select links that allow for southeast movement). White, Bernstein, and Kornhauser (2000) found that considering travel direction improved accuracy compared to a simple geometric approach. However, Zhou and Golledge (n.d.) concluded that those directional algorithms are problematic for GPS traces of slow-moving vehicles; therefore, such approaches may not work well for GPS traces of cyclists.

Topological methods improve network mapping compared to the geometric methods because they consider how network links connect together to limit selection to feasible paths. However, while topological approaches ensure route connectivity, the method still does not guarantee accurate selection of the path on the network because, for example, of the possibility of more than one feasible path or of selecting links in the beginning that result in a path that does not match well with the GPS trace in later segments of the selected path. Dalumpines and Scott (2011) developed an algorithm that incorporates geometric (i.e., proximity of GPS points to the links) and topological (network connectivity) methods that constrain the search for a path to avoid some of those pitfalls. That algorithm is described in more detail later, but in short, a buffer created around the GPS trace constrains the search for and selection of network links used to create the shortest feasible path within the buffer. The Dalumpines and Scott (2011) method allows for the consideration of multiple potential paths in a more simplified way compared to the advanced methods described in the next paragraph.

Several advanced methods have been developed that seek to eliminate the problems typical of geometric and topological methods. Pyo, Shin, and Sung (2001) developed the Multiple Hypothesis Technique that allows for several candidate routes to be kept in memory and assigned likelihood probabilities based on the position and heading of the GPS points and the topology of the road network, thereby hedging against the uncertainty inherent in the location of the GPS points. This method was extended by Marchal, Hackney, and Axhausen (2005), who added a topological aspect to the algorithm, and then again by Schuessler and Axhausen (2009) who refined the methods with link scoring and gap filling. Hood's (2010) network mapping of the CycleTracks GPS traces gathered in San Francisco, California, used the advanced approach developed by Schuessler and Axhausen (2009) that considers multiple criteria such as the angle of changes in the sequence of GPS points to determine whether a trip reached the end of a network link. Hood (2010) noted that the method only successfully mapped 1,454 traces from 260 participants to the network (out of 2,282 GPS traces), revealing the difficulties of finding an algorithm that will successfully map all the GPS traces to a network.

Methodology

This study prioritized selecting a network mapping method capable of being implemented in GIS because users interested in evaluating bicycle route data in the public and private sectors typically have access to and use GIS. The Dalumpines and Scott (2011) method described earlier and selected as the method to use for this research is a GIS-based algorithm that considers both the proximity of the GPS points to the network links and the feasibility of the path, therefore incorporating geometric and topological procedures. Dalumpines and Scott (2011) recognized that the GIS-based approach is less computationally efficient than the non-GIS-based approaches but they argue that computational efficiency is less critical when the data is being used for planning purposes rather than real-time applications.

Though the study in this chapter and the study conducted by Dalumpines and Scott (2011) use the same network mapping method, there are two major differences between the studies. First, the dataset used by Dalumpines and Scott (2011) consisted entirely of car commute trips, whereas this study consists of bike trips of a variety of trip purposes (e.g., commute, exercise, and shopping). Second, the Dalumpines and Scott (2011) study had access to travel diary data used for two main purposes: to verify the routes mapped to the network to see whether they matched the actual routes taken and to calibrate the user-specified buffer distance (used to constrain the search for paths) to obtain the best results. Dalumpines and Scott (2011) indicated their algorithm correctly matched 91 out of the 104 total car trips sampled in their study, for a success rate of 88 percent. Unlike their study, this study had only the GPS traces from the CycleTracks application and no other path data (e.g., a record of actual paths taken in travel diaries). Therefore, this study devised and used an alternate validation method to assess the accuracy of mapping the cyclists' GPS traces to the network (described in the results section of this chapter). The following section describes this study's design and application of the Dalumpines and Scott (2011) method.

Study design

From May 1 to October 31, 2011, researchers recruited cyclists in Austin, Texas, to download the CycleTracks smartphone application, which was developed by the San Francisco County Transportation Authority. Cyclists used the app to track their cycling trips and then uploaded their trips to a server where the research team could access the data. For extensive details about the study and participant recruitment, see Hudson et al. (2012). Over the course of our study, 3,264 GPS traces were collected from 316 participants.

Data cleaning

As mentioned in the Introduction, the first step in processing GPS data is data cleaning. Data problems included signal interruptions that can occur because of obstructions, such as heavy tree foliage or tall, multi-story buildings (Duncan and Mummery 2007). Figure 12.1 shows GPS traces in the downtown area of Austin with one type of problem commonly

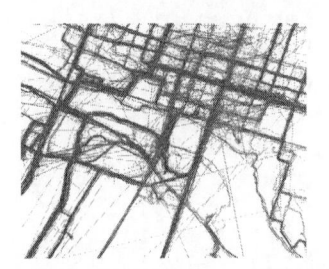

Figure 12.1 Map of pre-processed GPS traces showing errant GPS traces.

encountered in the data: segments of a route where outlying GPS points cause the GPS trace to "fly off" to another area.

For the first step of data cleaning, several columns were added to the GPS coordinates table to note the changes between each pair of GPS points collected. The columns were (a) distance traveled since last point captured, (b) change in time, (c) speed, (d) change in altitude, and (e) slope. Points within a trip were deleted if (a) the horizontal or vertical accuracy measurement was greater than 100 feet, or (b) the speed was greater than 30 miles per hour (not likely for a cyclist) or less than 2 miles per hour (at a stop or walking). Once those points were removed, the new column data (representing changes between points) were recalculated. To account for cyclist error in not turning off the app when not cycling or GPS noise, a customized code script split a single GPS trace into multiple separate traces if the time between points was three or more minutes or the distance between points was 1,000 feet or more. The code also removed GPS traces with fewer than five collected points from the dataset because those traces were considered too short to be actual cycling trips.

Network

Schuessler and Axhausen (2009) stress that an "important requirement for each map-matching algorithm to work properly is a correct, consistent, and complete representation of the real network by the network used for the map-matching. Unfortunately, hardly any network currently available can guarantee this requirement" (11). This limitation is especially the case with a network for mapping cycling paths because cyclists do not necessarily constrain their movement to the existing roadway network. Typical roadway network files do not include the paths taken by cyclists such as through parking lots and driveways, park trails, campus sidewalks, and open fields. Figure 12.2 shows an example of GPS traces through the parking lots

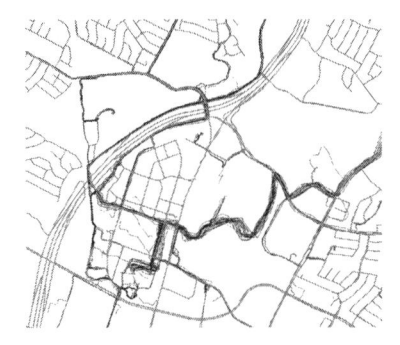

Figure 12.2 Bike trips traveling off the street network (between North Mopac and Burnet Road in Austin, Texas).

of office complexes in north Austin. Assigning the GPS traces to nearby existing roads instead of those alternative facilities would give the incorrect impression that cyclists use existing street facilities, when in fact the cyclists avoid the street facilities to shorten travel time, avoid unsafe conditions, or create a more pleasant travel route. To avoid losing information about the facilities used by cyclists, the time-consuming process of creating additional links in the original network (for this study, the City of Austin's freely available ArcGIS shapefile of the roadway network) was undertaken to create as complete a network as possible in the time provided. The network completion process resulted in researchers manually adding 923 additional links. Most of the additional links represent off-road paths through parks, parking lots, and driveways.

Mapping algorithm

Mapping the GPS points to the network follows the data cleaning and network completion steps. The literature usually refers to the last step as "map-matching"; however, that implies the GPS data is "matched" to the correct link. Since this study used data that does not have a corresponding travel diary or other data source to check the algorithm's ability to correctly match to the network, the step is referred to in this chapter as "mapping." As the first published algorithm to use ArcGIS for map-matching GPS traces to a street network, the Dalumpines and Scott (2011) algorithm provided a promising approach for this study. This algorithm, summarized in the following five steps, assumes no errors or gaps in the GPS traces (hence the data cleaning first step) and previous construction of a complete network dataset. Modifications to the algorithm for the research presented in this chapter occur in step three.

- *Step 1.* Connect GPS points for each bike trip into a line feature (i.e., the GPS trace).
- *Step 2.* Create a buffer around the line feature with a user-defined distance (see Figure 12.4). The buffer is used to delineate the line barrier that constrains the search for the shortest path.
- *Step 3.* Assign the origin and destination stops (the first and last GPS points) and the line barriers (the outline of the buffer) for the ArcGIS route analysis layer. The research presented in this chapter modifies the Dalumpines and Scott (2011) algorithm at this point to include additional stops between the start and end points in order to provide more information for the algorithm to map the GPS traces to the network.
- *Step 4.* Use ArcGIS's network analyst route solver tool to generate routes (i.e., paths) for each trip. The generated route is the shortest route (based on criteria such as travel distance) formed by the links contained within the buffer line barriers.

- *Step 5*. Depending on the network dataset used, update the trip attribute table to include the number of left and right turns along the constrained shortest-path route calculated during the route solver operations.

The ingenious aspect of the Dalumpines and Scott (2011) algorithm is their incorporation of ArcGIS's route solver tool that uses Dijkstra's shortest-path algorithm to find the shortest path between an origin and destination as well as the stops in between. Their algorithm constrains the search for the shortest path within a buffer width established a certain distance away from each of the GPS traces (see Figure 12.3). Dalumpines and Scott gave credit to Zhou and Golledge (n.d.) for mentioning the potential use of the shortest-path algorithm; however, Zhou and Golledge did not test the approach.

The tricky part of the Dalumpines and Scott (2011) algorithm is determining what buffer width to use. They found through sensitivity analysis of complex trips (e.g., contain loops or sharp curves) that for distances below 50 m (about 164 feet) on either side of the GPS trace the shortest-path algorithm did not find any routes; however, with buffer distances above 60 m (about 197 feet), some inaccurate routes resulted. Dalumpines and Scott (2011) recommended a buffer distance of 200 feet to maximize routes found (with the understanding that some may not be accurate). All of the algorithm experiments for the research in this chapter use the recommended 200-foot buffer width, such that the algorithm considers network links within 200 feet on both sides of the GPS trace.

For clarity, the variations of the Dalumpines and Scott (2011) algorithm tested in this study are referred to as DS2, DS3, and DS5. "DS" refers to Dalumpines and Scott (2011) and the number refers to the number of points (stops) that were used to constrain the mapped route. A "2" means that only the origin and destination were specified as stops the shortest path algorithm must reach (see Figure 12.4), "3" means that the midpoint of the GPS trace is also a stop constraint (see Figure 12.5), and "5" means that three intermediate points were used (the midpoint between the origin and destination, the point between the midpoint and the origin and between the midpoint and the destination) (see Figure 12.6). The use of three and five points took advantage of the ease of finding midpoints along a GPS trace using existing ArcGIS tools.

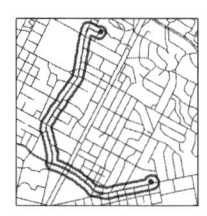

Figure 12.3 Buffer around GPS trace.

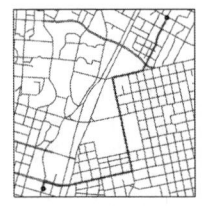

Figure 12.4 Example of route mapped by DS2.

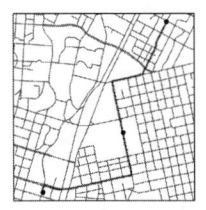

Figure 12.5 Example of route mapped by DS3.

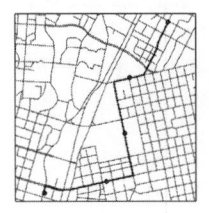

Figure 12.6 Example of route mapped by DS5.

Analysis of results

To assess the effectiveness of the DS2, DS3, and DS5 mapping algorithms, two types of analysis evaluated the results of the mapping process. The first considered how many of the individual bike trip GPS traces mapped to the network and assessed the accuracy of those mapped routes with a measure that compared the mapped route length with the original GPS trace length. The second type of analysis consisted of visually checking the mapped routes with the GPS traces.

As can be seen from Table 12.1, the percentage of GPS traces mapped to the network is essentially the same across the three variations of the DS algorithm. Runtimes for all the DS algorithms were about the same, regardless of the number of intermediate points. However, as Table 12.2 shows, differences emerge between the DS algorithms when trip purpose is considered. The percent of routes mapped is high for each trip purpose. However, accuracy varies, especially with exercise trips, particularly when using DS2.

Table 12.1 Overall results

	DS2	DS3	DS5
Total % mapped routes	87	87	86

Since this study does not benefit from having supplemental travel information to verify the accuracy of the mapping of the GPS points to the network, the accuracy reported is the percentage of mapped routes with a ratio of mapped route length to GPS trace length between 0.9 and 1.1 (in other words, mapped route length is within 10 percent of the original GPS trace length). The 10 percent was chosen simply to allow for some room for error due to the "noise" in the GPS data that may exist, even after cleaning. The accuracy percentage reported in Table 12.2 refers to the percentage of all the mapped routes for the trip purpose that met the criteria of being accurate.

Table 12.2 Results by trip purpose

	Average length (mi)	Standard deviation		DS2	DS3	DS5
Commute	4.36	3.44	Total # of Trips	1,418		
			Mapped (%)	86	86	86
			Accuracy (%)	77	77	76
Errands	2.65	2.77	Total # of Trips	313		
			Mapped (%)	85	85	85
			Accuracy (%)	64	65	65
Exercise	6.25	6.47	Total # of Trips	386		
			Mapped (%)	88	86	85
			Accuracy (%)	38	53	55
Other	2.64	2.86	Total # of Trips	102		
			Mapped (%)	89	87	87
			Accuracy (%)	65	68	66
School	1.88	2.86	Total # of Trips	144		
			Mapped (%)	89	89	89
			Accuracy (%)	65	66	65
Shopping	2.38	2.46	Total # of Trips	169		
			Mapped (%)	86	85	85
			Accuracy (%)	64	65	67
Social	2.81	2.99	Total # of Trips	381		
			Mapped (%)	93	92	92
			Accuracy (%)	71	72	73
Work	3.40	2.88	Total # of Trips	177		
			Mapped (%)	81	81	81
			Accuracy (%)	64	66	66

This method of determining accuracy has its limitations, of course, but provides an easy to calculate method of comparing the original GPS trace route with the resulting mapped route.

Route types

Conducting visual checks on selected routes to compare the GPS trace with the resulting mapped route helps further assess accuracy. Routes were selected for a visual check and the analysis reveals the influence of the number of points (stops) used in the DS algorithm and the differences that can be seen between the GPS trace and the resulting mapped route. The types of routes are: linear, loop, linear-loop, and mini-loop.

Linear routes

All three algorithms mapped simple linear routes (most commonly commute trips) that travel linearly from origin to destination without loops within the route accurately. From a check of the mapped routes, it was very rare to see a route that was matched by DS2 and DS3 and not by DS5. The traces by both DS3 and DS5 are usually very similar in majority of the routes unless the case is a loop route.

Loop routes

Figures 12.7, 12.8, and 12.9 illustrate, using one example route, the performance of DS2, DS3, and DS5, respectively, for loop routes. In Figure 12.7, the mapped DS2 route directly connects the origin and destination stops by a very short route that does not follow the loop at all. For loop routes, DS2 did not successfully find the used route. On every route sampled, DS3 backtracked after reaching the intermediate point (see Figure 12.8), and so only matched half of the route correctly. DS5 was the only algorithm that properly mapped the original loop GPS trace (see Figure 12.9).

Linear-loop routes

Linear-loop routes are defined as routes that follow the same path from start to mid-point as mid-point to end. As with the loop routes, DS2 found a very short path that connects the start and end points (see Figure 12.10 for an example). DS3 and DS5 both appeared to map the route accurately (see Figure 12.11 and Figure 12.12).

Routes with mini-loops

Mini-loops are defined as small loops within a route. All of the algorithms tested had a difficult time accurately matching routes with mini-loops, since

Figure 12.7 Application of DS2 to a loop route.

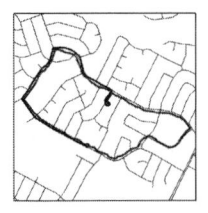

Figure 12.8 Application of DS3 to a loop route.

Figure 12.9 Application of DS5 to a loop route.

even five points is not sufficient to properly trace a route this complicated (see Figure 12.13 for an example).

Conclusions and recommendations

GPS-enabled smartphones produce data useful for tracking cyclists' routes, but because of the inherent errors in tracking movement with GPS, analyzing the data requires an algorithm to map routes to the established network of roadways and off-road paths. Using a modified ArcGIS-based algorithm developed by Dalumpines and Scott (2011), this chapter makes the following conclusions and recommendations.

For best results and to maximize success of mapping GPS traces to a network, GPS data must go through a robust cleaning process to remove GPS "noise" and the network file will most likely need to have additional links added to fill in possible gaps in the network, such as links to represent off-road paths through parks, parking lots, and fields.

Figure 12.10 Application of DS2 to a linear loop route.

Figure 12.11 Application of DS3 to a linear loop route.

Figure 12.12 Application of DS5 to a linear loop route.

Figure 12.13 Application of DS5 to a route with mini-loop.

The success of mapping the GPS trace to the network also requires consideration of the shape of the trace (e.g., linear, loop, linear-loop, and mini-loop). The Dalumpines and Scott (2011) algorithm that uses two points (DS2, using only the origin and destination points) maps the linear (typically commuter) cycling trips well if the paths are relatively straight.

However, for GPS traces that consist of loops (e.g., the origin and destination are the same and the cyclist is not repeating the same route out and back), at least two intermediate points are needed to map the GPS trace to the network to avoid the algorithm mapping the trace to the midpoint of the trace and then returning back to the origin without proceeding to the other half of the loop. Since run times are approximately the same for all variations of the DS algorithms, the DS5 (three points between the origin and destination points) offers protection from incomplete/inaccurate mapping of a trace if trip purposes (e.g., exercise, commute) are unknown and visually checking all traces for type of path (e.g., linear and loop) is too cumbersome. However, as the number of intermediate points increases, the constraints on the algorithm increase and may lead to GPS traces not being mapped to the network. A methodology that involves using all three variations of the DS algorithm (DS2, DS3, and DS5) on the same dataset and identifying the cycling trips with different paths from each variation that need to be visually checked can overcome those issues.

In addition to demonstrating use of the DS algorithm variations, this chapter presented the application of two analyses to assess the accuracy of the paths mapped to the network from the GPS traces. Though the analysis requiring a visual check of resulting routes can be time-consuming, the metric of relating the length of the mapped path to the length of the GPS trace provides an efficient means to assess reasonableness of results and to quickly identify outlier results that need further examination. Of course, accuracy cannot be known for sure without another source to confirm the path taken by the cyclist, but with the cost, time, and inconvenience of requesting travel diaries or interviews, the accuracy assessments described in this paper provide a more affordable and efficient alternative.

Ideally, a smartphone application that maps the path in real time during the cycling trip and then asks the cyclist to confirm accuracy of the mapped path at the conclusion of the trip (and revised as needed) would provide the best source of information. Until that is developed, in wide use, and results available to public and private sector transportation analysts, the methods presented in this paper provide a means to extract, as accurately as possible, paths taken by cyclists from GPS traces collected from smartphones.

Acknowledgments

This work was funded partially by Texas A&M University's University Transportation Center for Mobility and was also partially supported by funding under an award with the U.S. Department of Housing and Urban Development. The substance and findings of the work are dedicated to the public. The authors and publisher are solely responsible for the accuracy of the statements and interpretations contained in this publication. Such interpretations do not necessarily reflect the views of the government. The authors would also like to express appreciation to staff at Esri for assistance with the network mapping, and the San Francisco County Transportation Authority for providing us with access to the data and the CycleTracks application.

References

Dalumpines, Ron and Darren M. Scott. 2011. "GIS-Based Map-Matching: Development and Demonstration of a Post-Processing Map-Matching Algorithm for Transportation Research." In *Advancing Geoinformation Science for a Changing World*, edited by Stan Geertman, Wolfgang Reinhardt, and Fred Toppen, 101–120. New York: Springer.

Duncan, Mitch J., and W. Kerry Mummery. 2007. "GIS or GPS?:A Comparison of Two Methods for Assessing Route Taken During Active Transport." *American Journal of Preventative Medicine* 33(1): 51–53.

Hood, Jeffrey N. 2010. *A GPS-Based Bicycle Route Choice Model for San Francisco, California*. Professional Report, University of California, Berkeley, CA.

Hudson, Joan G., Jennifer C. Duthie, Y.K. Rathod, Katie A. Larsen, and Joel L. Meyer. 2012. *Using Smartphones to Collect Bicycle Travel Data in Texas*, Report No. UTCM 11–35–69 for U.S. Department of Transportation.

Marchal, Fabrice, Jeremy Hackney, and Kay W. Axhausen. 2005. "Efficient Map Matching of Large Global Positioning System Data Sets: Tests on Speed-Monitoring Experiment in Zurich." *Transportation Research Record: Journal of the Transportation Research Board* 1935: 93–100.

Newsom, Paul and John Krumm. 2009. *Hidden Markov Map Matching Through Noise and Sparseness*. ACM GIS 2009. Seattle, WA.

Nielsen, Otto Anker, C. Wurtz, and R. M. Jorgensen. 2004. "Improved Map-Matching Algorithms for GPS Data Methodology and Test On Data from the AKTA Road Pricing Experience in Copenhagen." *19th European Conference for ESRI Users*. Copenhagen.

Pyo, Jong-Sun, Dong-Ho Shin, and Tae-Kyung Sung. 2001. "Development of a Map Matching Method Using the Multiple Hypothesis Technique." *2001 IEEE Intelligent Transportation Systems Conference Proceedings*, Oakland, CA.

Quddus, Mohammed A., Washington Y. Ochieng, and Robert B. Noland. 2007. "Current Map-Matching Algorithms for Transport Applications: State-of-the-Art and Future Research Directions." *Transportation Research* Part C, 15: 312–328.

Schuessler, Nadine and Kay W. Axhausen. 2009. *Map-Matching of GPS Traces on High-Resolution Navigation Networks Using the Multiple Hypothesis Technique*. Working paper. Zurich: Swiss Federal Institute of Technology.

White, Christopher E., David Bernstein, and Alain L. Kornhauser. 2000. "Some Map Matching Algorithms for Personal Navigation Assistants." *Transportation Research* Part C, 8: 91–108.

Zhou, Jianyu and Reginald Golledge. n.d. "A Three-Step General Map Matching Method in the GIS Environment: Travel/Transportation Study Perspective." Berkeley, CA: University of California Transportation Center. http://uctc.berkeley.edu/research/papers/767.pdf.

Index